Thomas Hughes

Principles of Anthropology and Biology

Second Edition

Thomas Hughes

Principles of Anthropology and Biology
Second Edition

ISBN/EAN: 9783337216481

Printed in Europe, USA, Canada, Australia, Japan

Cover: Foto ©berggeist007 / pixelio.de

More available books at **www.hansebooks.com**

PRINCIPLES

OF

ANTHROPOLOGY

AND

BIOLOGY.

BY

REV. THOMAS HUGHES, S.J.

SECOND EDITION.

NEW YORK, CINCINNATI, CHICAGO

BENZIGER BROTHERS,

Printers to the Holy Apostolic See.

1890.

PREFACE.

———

This reproduction of four lectures, delivered before the members of the Detroit College Alumni
Association, and published by the same gentlemen
during the winter of 1888–89, is respectfully dedi-
cated to the cultured classes of the community,
and to the advanced students in colleges and acad-
emies, who frequently ask what line of theoretic
truth is to be followed, in the midst of so much
scientific research.   If we would not lose the best
part of the practical results which science offers us,
we must keep scrupulously to the line of truth
which sound logic requires us to follow.

THE AUTHOR.

New York, July 31st, 1890.

CONTENTS.

ANTHROPOLOGY.

CHAPTER I.

PREHISTORIC RACES.

(N. B. The numbers refer to the paragraphs.)

The prehistoric difficulty, 3 ; barbarism, 6 ; geography of the prehistoric, 10 ; methods of induction, 13 ; geological chronology, 15 ; argument of induction, 16.

Archæology, 18 ; ages of metal, 19 ; neolithic age of polished stone, 25 ; evidence, 28 ; palæolithic age of chipped stone, 29 ; glacial epoch, 30 ; epochs, periods, formations, 31 ; results of archæology, 32.

Palæontology, 33 ; extinction of species, 34 ; subdivisions of the ages, 35.

Anthropology, 36 ; tertiary man, 38 ; man and geology, 46 ; man and the universe, 47.

CHAPTER II.

ACTUAL RACES IN HISTORY.

The term, species, 51 ; the test of species, 52 ; the term, race, 55 ; analogies of the lower orders, 56 ; variations, 58 ; argument of analogy, 59.

BIOLOGY.

CHAPTER III.

SPECIES ; OR, DARWINISM.

CHAPTER IV.

CELLS ; OR, EVOLUTION.

ANTHROPOLOGY.

CHAPTER I.

PREHISTORIC RACES.

1. In the congress of German naturalists and physicians held at Wiesbaden, the celebrated Berlin professor, Virchow, delivered an address on the progress of anthropology and biology. Reviewing, under the double aspect of prehistoric and historic man, the present state of anthropology, he expounded several views at considerable length. Anthropology is the science which treats of the human species in its natural groups and general formation. It involves the study of all human characteristics, as well physical, physiological and pathological, as moral, social, and political. The professor stated that, as to prehistoric anthropology, every positive advance which we had made in that study had removed us farther than before from any proof of evolution to be found there. Man has not arisen from the ape, nor has any ape-man existed linking the two species together. Then, as to historic races, he proved that the Australian, which is quoted as being the most imperfect among them, is shown to

9

be nowise ape-like, but entirely human, like ourselves. Finally, touching the biological question of the transformation of species, he affirmed that it is not yet possible to give any certain proofs of man's tertiary origin in the world.

2. Such a statement of the question, coming from such a quarter, seems to be a propitious augury that the great fight of evolution, after lasting for more than thirty years, is, like other wars of long duration, approaching a final issue. Many signs of the same coming event have been discerned elsewhere. And, taking this state of things as our point of departure, we may review the anthropological question, as it has stood thus far, and as it seems to be nearing its solution to-day. Subsequently we shall take up the question in biology. The manner of treatment which recommends itself is not that of the specialist, but that of the philosophical critic, who gauges the value of scientific proofs by the general laws of reason and philosophy. Leaving, then, the German and other specialists aside, it is with this school of criticism that we venture to range ourselves.

———◆———

3. To apprehend the prehistoric difficulty which attaches itself to anthropology, I would invite you

The Prehis-
toric Diffi-
culty.

to take a stand upon some commanding spot, whence the whole field of the contention may be surveyed. We can thus conceive, too, some preliminary notions on the pre-

historic and savage states of humanity. Perhaps
no better position offers itself than this present
moment of time which is now passing, and this
point of space in which we now happen to be. We
are here, and now : we are defined by this moment
and this point. How different has it been with the
family to which we belong! Men and women have
lived, and their hearts have throbbed, all over the
habitable space on this globe of ours, and all the
way back through the ages past, in places where we
have never set a foot, in ages long before we were
born. The course of our family's history, originating in a definite place and at a definite time, has
flowed outward and onward to all the borders and
limits of this habitable globe. Unlike the fated
sameness of any dumb species of animals, with its
instincts running in a fixed channel, and the expression of its life about as rigid as a scientific formula, the story of our family has been rather that
of a turbulent sea, swelling and surging in all directions. It has ever had a free and self-willed nature.
It has ever been a restless body of vitality, just
kept within some bounds of time and space and
eternal laws, by one Power which knows how to
limit the tide.

4. Defining in particular the position of our ancestry, with reference to the knowledge
that posterity were to acquire of them,   **Outside of History.**
we may note that different fortunes
have attended different lines in the family's an-

tecedents.   Some parts of its eventful course have
been happy enough to find historians, and have been
described in the faithful reports of men living, observ-
ing and writing at the time that events happened, or
within a reasonable and speaking distance of men
who lived at the time.   Such reports give us what is
called documentary or monumental history.   But
there are parts also which are prior to certain lines
of documentary history, or which lie, some way or
other, outside of the margins of any local records.
They are like the portions which the Chinese comprise in their annals, but which they expressly designate "parts outside of history."   Such unrecorded
antecedents of our history the present age has been
pleased to call " the prehistoric."

5. Thus in France, Denmark and England, in
America, North and South, we may
Outside of Civilization. discern with the aid of archæology the
tidal remains of an ancient humanity,
which must have welled up from its primal springs
somewhere, probably in the East, but thence over-
flowed and rolled on to what were as yet but vacant
shores.   Whether such relics are to be found in
China and Japan, we are not yet informed.   As
that overflowing population rolled so far away from
its origin and its source, it lost in many instances
the best part of its civilization, just as we should
lose it now, with all our culture, nay, because of
our delicate culture, if we were stranded on barren
islands.   It lost its social depth, and carried with

it but the fractured relics of facts, traditions and histories, of arts and crafts, and even of the very means and methods of subsistence. If, then, in the lands of its colonization, the dire evil of famine, and the intense cold of an age of ice overtook it, what else should we expect but to find it in holes and dens, with the bear and the deer in its very midst, destitute and soon degraded, in numbers few, like the Esquimaux or Alaskans, and with families extremely small? This last circumstance, of a limited offspring, seems to follow in such surroundings, either because of other reasons that we might think of, or, as the latest official report from Alaska mentions, simply because of the hardships of their condition. And what hardships those are, when the wife so inevitably becomes a mere drudge, a slave, even to her youngest sons! On these terms a few generations would plunge the most civilized of us into barbarism.

6. Barbarism is a state of things which results from the composition of two factors, human beings to become destitute, and desperate conditions of life to make them so. There **Barbarism.** is plenty of room to imagine well-nigh desperate conditions of existence, and therefore impossible conditions of civilization. The caves and holes of an icy cold age, with wild beasts prowling about, and, instead of lending us their skins to keep us warm, choosing rather to make their meals on us, and on our children—these and other such inter-

esting situations, which prehistoric archæology, as we shall see, quite significantly suggests, would reduce the best of us to the abject condition of "cave men," taken at their worst. And, possibly, if there was nothing better to be had, we might reconcile ourselves to things as they were; especially when, all distinct recollection of a better state dying away in the course of time, custom with its strong, nervous bonds of a second nature could give men a positive preference for a cave or a hole, as we know it gives some a preference for a craggy hill-top or a smoky tent. Thus, in fact, we see that troglodytes, or men who live in caves, are recorded all through history.

7. Not a little rhetoric has been expended on the savagery of these cave-men, and the origin which

Cave-men. must have been theirs down among the tribes of apes. So it is worth our while to observe that, on the contrary, the more civilized the men had been before, that is to say, the more resources they had enjoyed outside of themselves for procuring food, clothing, and shelter, the fewer resources then would they find in themselves, and the more abject would their condition be, in the circumstances which we are contemplating. We may bring this matter home to ourselves ; for it is quite possible that the present civilization will collapse into depths undreamed of now. Other great civilizations have vanished like a dream of the night before us. And what we say is this, that in a similar contingency, starvation would follow for ourselves and our posterity. Add, then, to the physical conditions which are always within easy distance of realization, as geologists, astronomers, and physiologists of the sea can tell us—add the moral conditions so soon to follow, of rapine, cruelty, and the other vices attendant on a collapsing state of society. Why, with all the terrors of menacing war and civilized control around, how hard is it to keep in check the brutal element of human nature, either in a country at large or in a single great city! One is reminded of the story how Adonibezec fled from the battle-field, and they pursued him and took him, and cut off his fingers and toes. And what did the wretched man say? " Seventy kings," he said, " having their fingers and toes cut off, have gathered up the leavings of meat under my table!" What must it be when civil authority is no more, martial law has no terrors to display, traditions are dying out, religion breaking up into idolatry, every man's hand against his neighbor, and all ready to pounce upon the weakest! Such individualism issues in barbarism, yes, African degradation, cave-men, troglodytes, almost ape-men. But then the ape-men will have come down from above; they will not have mounted up from below!

8. And when out of chaos order does arise again, owing to the infusion of a new blood, or to some genius actuating the potential vigor of human

nature, human still in the midst of its degradation, yet states and periods, republics and empires
Periods never have no such resurrection before them.
recur. They have only a single course to
run, a single goal to reach and turn, and, fleeting
like a courier, they are seen no more. They live,
grow, and dissolve; there is no resurrection for
them. So that, if the records are not saved before
the courier disappears, he will never return to bring
them. All will have faded into the prehistoric.

9. The obscurity enveloping such a movement
of transition, when barbarism is one of the termini,
has given some wide scope to certain
Prehistoric platitudes about these cave-men. The
Platitudes. air of an ascertained geography and
chronology is thrown about these ancestors of ours,
who are to be conceived, it is said, as crouching in
caves and crunching the bones of wild beasts. A
specimen of such platitudes offended our eyes the
other day, when, answering the ex-Premier of England, a noted writer discoursed some rhetoric thus,
in his most conclusive style : " It is hardly possible
to conceive of the years that lie between the caves
in which crouched our native ancestors crunching
the bones of wild beasts, and the home of the civilized man. Think of the billowed years that must
have rolled between these shores!" Here is an air
of scientific geography and an immeasurable chronology thrown about poor people, who certainly were
badly off. But it scarcely requires science to see

that a short time can suffice to drag people down; and not quite an interminable time is needed to lift them up again. The great and active energies in human nature are only waiting for the right touch and pressure to yield up their resources for use and development. How it has been with them in the past, we shall see, when we have determined the facts of the case.

This is enough, then, of the preliminary notions, which will serve to fix the scientific imagination on the study before us. Now let us address ourselves to the facts of the case, and see what interpretation they call for, and will bear.

10. We may first sketch the outlines of the geography which the prehistoric has really covered. Then the chronology will come in order, when we consider the ages, as they are called, of iron, bronze and stone. Pray observe Geography of the Prehis-
that the geography here will reproduce, toric. in its distribution of human fortunes, some of the same social phenomena which we witness on a smaller scale in the mixed population of any great city. There you may find opposite extremes at the same time of penury and opulence, within a stone's throw of one another, separated by just a street or two on the right hand or the left. And here, in the geography before us, you will find the prehistoric separated from the historic only by a natural boundary, as the Alps or the Danube; such barriers as have always been enough to sepa-

rate one race from another, and keep both unmixed. Or else the prehistoric dissolves into the historic, on the same ground; as in America, which is certainly now the subject of luminous history, we have only to go back four hundred years and we reach the line and cross it, and we are away in dim prehistoric America. So too is it with the greater part of Africa. In Iceland, Britain, Gaul, and Germany, in the lands of the North-East, overflowing with that population of Hun and Goth which poured into Europe, we have to travel back not two thousand years and we are stranded on the shores of the pre-historic. All the while, during four and six thou-sand years, other lands are abounding in monuments, written records, trustworthy traditions. But, over the ground of the prehistoric, records are wanting ; the induction of science alone is available; and we interpret as best we can the relics of archæology, of palæontology, of anthropology, which have escaped the ravages of time; and we note the few geological touches which the same ravages of time have left behind them. These geological data are few, for the records of the rocks were far on toward com-pleting the last chapter of their history, when man with his hopes and his fears entered on this arena of his short and anxious career.

11. Some lands seem never to have had a pre-Lauds never historic humanity to grace them or to
Prehistoric. blight them. All humanity there, how-ever ancient, is in the full light of history.

Beyond it there gleams a dawn of mythology and fable—but not the fable of an ape-man coming up from the tribes of brutes; quite the opposite, gods coming down to be heroes and men. Nor is there any reason to believe that these are the youngest of the nations; rather they are the oldest. And some do not recede even into the twilight of mythology; they are historic back to the very first, scientifically and critically historic, if documentary and monumental records have any value upon earth. Cradle-lands such as these never had anything to do with prehistoric races, except as bordering on them, or as originating them, in the sense which we shall explain farther on; inasmuch as these are the lands which sent forth such races on their melancholy wanderings, till the day should come when, dead or alive, their relics would be revealed in the far-off history of the future, and become the subject of a science, anthropology, yet to be. Alive, such relics appear in Australia and in the land of the Bushman; dead, in the extinct races of Canstadt, Cro-Magnon and others.

12. The nations that stayed at home were those of Egypt, Babylonia, Arabia, Persia, Phœnicia, India. They were stationary in more Nations that senses than one. They stayed at home, never retro-and they were conservative besides. graded. In consequence, they never lost so much that they reached a state of savagery, or that they ever had a journey to take back toward civilization. It may

be said of them that, if they are not progressive peoples, one reason is this: they never retrograded so far as to become nations of progress under the spur of reaction. Had they fallen lower, they might now stand higher. But their immobility forbade progress. The conservatism of these sons of Sem is not that recuperative power which the sons of Japhet have, and which the sons of Ham conspicuously have not. Yet they must have lost something under the friction of ages. Their conservatism could not guarantee them against the wear and tear of time. Hence that very immobility of theirs, so proverbial in history, serves this excellent purpose of showing that since they never gained anything, for it was not in them, and yet they must have lost not a little, for that is the condition of all things human, they are a standing monument of people who could not have come up from a state of savagery to be what they are to-day. They, and the rest of us too, have come down from a state of higher civilization. It is so easy a matter to run down, as every organism and every mechanism shows us! The whole history of our family comes to this: it has done best when it kept what it had, and next to best when it got back what it had lost. We have yet to find the nation which, without the help of revealed religion, shows signs of having still as much as the family, by all accounts, possessed at its origin. Some have never declined much, some have declined to rise again, others never to rise again; and all alike show that we have never risen higher than our origin, and those amongst us have done best who have kept nearest to the level of that.

13. Let this suffice to sketch the geographical outlines of our subject. We may now state the methods which science adopts to form its inductive conclusions. It endeavors to find all the traces possible of human life prior to historic times. It deciphers and interprets such traces, as indicative of the physical, intellectual or social condition of the men who existed then. These traces and indications of antecedent human life are to be found in the nature of certain objects imbedded in the soil, or otherwise preserved; they are also deciphered in the location, situation, where such objects are met with undisturbed.

Methods of Induction.

14. The objects in question are, first, the fossil relics of men themselves. These appertain to anthropology proper. Secondly, they are the fossil relics of animals that lived with men; and these pertain to palæontology. In the third place, there are weapons and utensils which men made and used —articles of industry. The degree of perfection or imperfection discernible in their make reflects upon the degree of civilization which produced them. For the material out of which they are made may have been easier or harder to procure, as stone is easier to get, or a bone, than bronze or iron. Or the

work expended on the material may be better or
worse, as chipping the stone is an inferior process
to smoothing and polishing it.   All this is the sub-
ject-matter of archæology.   Finally, as to the loca-
tions, these objects, of whatever kind or workmanship, have been preserved for us, as a general rule,
by the successive deposits of soil covering them in
the course of time, and thereby "fossilizing" them;
that is to say, putting them in the condition that
we have to dig them up, or unearth them now.
Deposits of soil, or stratification by natural agents,
are referred to geology, which thus is called upon
to interpret the antiquity of those strata, and therefore to settle the antiquity and chronology of prehistoric man, whose relics are found in the strata.

15. We may estimate at once the value of this
geological chronology, this determination of time
by the computations of geology.
**Geological
Chronology.** Whether this science is engaged in contemplating the last sediment deposited
right under our eyes in the Mississippi, or in com-
puting the time required for the entire formation of
the terrestrial globe, it cannot be credited with the
qualifications of an exact time-keeper; nor in its
practical efforts, when tested by actual observation,
has it come out felicitously in its results.   The reason seems to be that it has a time of its own, indeed; but geological time is not our historic time;
and there is no ascertained formula to make the
reduction of one in terms of the other.   The meas-

ure which natural forces employ in laying down a stratum of clay on the bed of a river is certainly not the same as the measure which divides the story of mankind into days and years and centuries, by solar and terrestrial revolutions. Nor is it the same as that which marks off generation from generation among mankind, dividing them by the births and deaths of men. Geology, in the order and in the thickness of its deposits, agrees with neither of these processes, neither that of astronomy nor that of anthropology. If the astronomical revolutions have been regular, if the revolving cycles of human generations have been quite irregular, the evolution of the earth's present surface has had a measure of its own, a time of its own, not identical with either of theirs. Yet, to derive any light from geology on the subject of man's antiquity, its time must be made commensurate with man's time. And accordingly the inductive effort has been made to argue from what we see, in present circumstances, going on in certain places that we know, to what we have not seen, in circumstances and places entirely different and unknown. But this inductive effort is faulty, because there is no induction about it.

16. Induction, as a form of argument, requires a sufficient enumeration of phenomena to formulate a general law, which is found **Argument of Induction.** to stand the test of verification on being applied to cases known, and which, therefore,

may be relied on to interpret things rightly when applied to the unknown. Here there is no sufficient enumeration of facts to formulate any law. Solitary facts known are compared with solitary unknowns; and conclusions are jumped at from such premises as logic will not admit, to begin with. And the end of the argument corresponds to the beginning. For when the hastily constructed law, derived from a few known facts, is tested in actual cases under our eyes, it is found so often faulty in its sum total of years required to fossilize a tree on the bank of the Mississippi, or to lay down ten or twenty feet of loam, that, whenever geology pretends to measure its time in terms of history, we are perfectly justified in suspending our judgment, until it has found a common denominator for historical duration and its own duration—two very different things.

17. It does mark the order of succession, whether in the soils deposited, or in the objects
Geology
Inductive. which those deposits contain. It marks too the relative proportion of duration, which respective thicknesses of the stratification seem to have required. But, with all that, the conditions of earth, and water, and air, and sky have been so different at the different periods of terrestrial evolution that, to read the lesson of stratification aright, there would seem to be needed an equipment of science on pretty nearly all the laws of the universe. Astronomy, meteorology, geogra-

phy are referred to in explaining geological formations; natural physics and terrestrial physics; min-
eralogy and chemistry; botany, zoölogy, physiology;
comparative anatomy. Geology, in fact, is a science of induction which bases itself on all the orders of facts and on all the laws in the boundless field
of nature.

No doubt, within the restricted limits of the present question, that of the prehistoric antiquity of
man, it does not lie open to all these uncertainties,
because it does not appeal to so many exact sciences. Still, not being exempt from a limited sum
of the scientific references, it remains liable to a
moderate sum of the consequent uncertainties. In
brief, geology is not the science to arrange an exact
chronology for the prehistoric periods. Let us see
if archæology has done so, or palæontology, or
anthropology, strictly so called.

18. The archæological results are as follows.
Prehistoric articles of industry have been found in
great numbers; and numerous, too, are
the localities in which they have been Archæology.
unearthed. There are stones, and bronzes, and copper; tools, chips and flints; there are places called
Danish *kjökken-möddings*, and there are Swiss lake-
dwellings; besides old hearths and camping-grounds,
and caves and other holes in the earth.

19. The reports from these and about them are
summed up in the theory of what are called the
ages of iron, bronze, copper, stone. Supposing

ourselves to be at present, as is evident, in an age

Ages of Metal. of iron, we must go backward to where written or historic annals began to be dated over the greater part of Europe; and we come upon a prehistoric time, in which also it is found that iron was used. This is, therefore, reckoned a prior age of iron to our own, and is otherwise called the Halstattian age. In this most recent prehistoric period, various specimens of iron work and great swords of iron are found; also special types of work in bronze. There are ornaments and razors in bronze, vases and other objects done in Tuscan style. It is the time of burying under mounds, whether with or without cremation.

20. These iron implements, which are taken to denote an earlier age of iron, are found in England, France, Spain, Switzerland, Denmark, Sweden and Norway, Italy; in America; in Egypt, Western Asia, New Caledonia, the Isles of the Ocean. Bronze relics, which some take to mark an age of bronze, are found in most of the same European countries, in Egypt besides, as also in Mexico and Peru. Copper objects too are met with in North America.

21. There is a grave objection to this idea of an

Objection to the Ages of Metal. age being prehistoric, in any sense to suit the purpose of evolution, if metal was used at all and worked as a material of industry. Metal is used in all modern times, and requires advanced workmanship. Stone, as

being easier to grind or to chip, might antece-
dently be regarded as the material for savage days.
But we shall see that even stone has its uses at all
times, within historic limits too; and, upon occasions, seems to be preferred. What ground can
there be for dividing off a prehistoric age of metal?

22. If the ground is this—a preconceived theory
that metallurgy, or the working of metals, must be
found somewhere in an incipient and transitional
state, following on a supposed earlier age of ignorance and ape-like incapacity, we have only to remark that such a latent theory is, in the first place,
a gratuitous postulate, assuming the very thing to
be proved, if evolution is to be made to stand. In
the second place, it is invalidated or contradicted
by scientific and documentary evidence. For scientific explorations in Egypt, Assyria, Babylonia,
show the use of metals there so far back in the past
that there is no warrant as yet for affirming the existence of a previous age, either of stone or of anything else. And positive documentary history informs us that, in Asia, tools of bronze and iron were
a product of industry as far back as Tubal-Cain, very
long, indeed, before historic annals began to be
dated over Europe.

23. As to bronze, in particular, authors consider
themselves qualified to deny entirely
that such an age existed anywhere. **Bronze.**
Perhaps, however, in a modified sense, that may be
called an age of bronze in Switzerland and Savoy

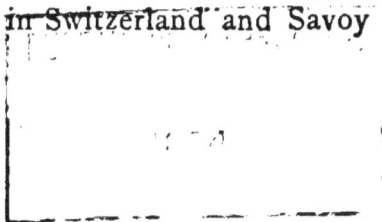

which witnessed a great local development of this kind of metallurgy, as the rich stations of bronze-remains in those places seem to indicate. Or, at times, it would appear that bronze was specially imported into a certain country, as Egypt; or, generally, becoming an article of commercial importance, it overspread Europe with Tuscan work.

24. But if this is all that the " age of bronze " comes to, or the " age of iron," then the analysis of history is making these prehistoric periods less distinct, the more we know. It looks as if this romance of a scientific generation were coming down to the homely synthesis of Moses, Homer and Livy, and as if the imposing term " prehistoric," which has been so vaguely magnificent in science, because so magnificently vague, were but a new phrase for the old idea, " Once upon a time!" equally obscure, but less pretentious.

25. Older than the age of the metals is that of stone, and, first, of polished or ground stone. This is otherwise called the

Neolithic Age.

neolithic, or newer stone age; the one to be mentioned next being called palæolithic, or older stone period. The specimens of work which are referred to it are axes, chisels, etc., made of such materials as diorite, serpentine, basalt, quartzite. There are clay vessels, too, hand-made but elegant. In the artificial shell-deposits which are seen in Denmark, and are referred to this epoch, there are found tools of flint, horn and bone, fragments of a rude kind of pottery, charcoal and ashes, but no objects of metal. The earlier great stone, or megalithic monuments of Europe, called dolmens, or chambered tumuli, belong to this period.

26. The men who lived then were not mere hunters; they were tillers of the soil. The bearing of this distinction will appear subsequently. Some are inclined to believe that the polished stone period was inaugurated in Europe by the spreading of a new population, in which they would recognize the first wave of Aryan immigration. That there was a sudden infusion of some new people is rendered plausible by the gap which is found in the process of transition from the implements of an earlier age to those which characterize this one: there is a want of intermediate forms to mark what might be considered stages of evolution. Besides, there is noticed the presence now of divers species of domesticated animals, presumably brought by a new people from distant countries.

27. The lands, in which this neolithic age is discerned, are the same European countries enumerated before for the age of iron; Coeval with Rome. along with North and South America, Terra del Fuego, Australia, New Caledonia, South Africa, the Isles of the Ocean. The oldest lake settlements of Switzerland belong to the same period. Many, however, of the settlements in the most Western Swiss lakes must have been flourishing rather late in history. For what do we find?

There come to light various articles of bronze, weapons of iron, and even coins of Roman origin. Does this mean that the prehistoric age of polished stone is coeval with Roman history? It begins to appear that the prehistoric man of even the stone age was a man that trafficked, and perhaps fought, with the dread legionaries of the Roman republic, or perhaps the Roman empire. In fact, it dawns upon us, as scientific investigation advances from hazy theory into the broad light of ascertained results, that he is only our well-known European cousin, or American, or Australasian, studied by other lights than those of written history; whom we have met often enough under a different name from the " prehistoric " in the pages of his contemporaries, Herodotus, Livy, and even Tacitus, or writers later still. The dimness of our view in this study is owing to the fact that we are looking at him under the light of inductive or inferential evidence, not that of palpable observation, or documentary record.

28. We may note, in passing, the different kinds of evidence that may be brought to bear on a subject. There is written history, which furnishes documentary evidence: this is the chief means of knowing our human family. There is pure theory, which carries with it a kind of speculative light, to show the possibility of things being true. There is natural science, which proceeds by way of direct observation and experiment; and this

Evidence.

gives us evidence which is immediate and conclusive; but its application is limited to our surroundings, and cannot reach into the past or future, or
to things distant. To reach those things which do
not fall under observation or experiment, natural
science infers from what it does observe, from the
data which are thus supplied; but the evidence becomes then only indirect and inferential. If any
elements of mere theory or hypothesis are now
added to the inferential process, evidence ceases
and we have probability instead; and the final conclusion partakes more and more of theoretic probability, or even bare possibility, according as more
elements of theory are inserted in the premises.
And, if an ingredient of false theory is anywhere
added, the final precipitate of the compound process will be anything but the truth. It may be an
agreeable doctrine, popular, fair to see; but not true.

29. Older than the neolithic age, with its man of
the ground and polished stone, is the
palæolithic or ancient stone age, with   **Palæolithic Age.**
its man of the chipped or flaked stone.
He helped himself to what utensils or weapons he
needed, by chipping rude stones into some shape or
other of axes, lance-heads, or the like. The man of
this time was probably a hunter or warrior, the vanguard of coming immigration. He was overtaken,
apparently, by a period of such intense cold, that
it reduced the greater part of Europe to the conditions of an Arctic climate.

30. Among the theories devised to account for this
Glacial cold period or glacial age, as to which it
Epoch. is still dubious whether there was only
one such spell, or more than one, the latest hypothesis, that of an Italian philosopher, connects
it with the deluge, a phenomenon reported to
us with the most exact documentary evidence.
According to this theory, the glacial period coin-
cided with the flood, in the sense that the reign of
ice was brought on by the causes which operated
during or after the deluge.   So that the deluge
would have to be conceived as an event, which,
either before or after or concomitantly, involved a
revolution in all parts of the globe.   And the part
which we know of so well, as described by eye-wit-
nesses in the Mosaic narrative, would then be
merely one phase, one scene in a tragedy very
great, one episode in a terrible drama that involved
the whole of our orb.   How the palæolithic man
dipped into the glacial age in Europe, we do not
see distinctly stated; whether it was that, after the
deluge, roaming far away from the cradle-lands of
the family, he found himself in places suffering
from this Arctic cold, and he became hopelessly
ice-bound there; or that, according to another
most recent speculation, the value of which is not
yet determined, the flood of overflowing waters did
not actually reach all parts of the earth; and he,
in his own home, was enveloped in some of its marginal phenomena, among which was this intense

atmospheric cold. All these, however, are specimens of purely theoretic probabilities so far; the
major part of the light to illumine them being nothing more than speculation.

31. Before the glacial period, which bound the
zone of temperate Europe in fetters of **Epochs,**
ice, the climate in the same parts was **Periods,**
most mild, and even tropical. The fos- **Formations.**
sils of the tertiary age, which had just elapsed, exhibit palms, cypresses, plane-trees, fig-trees, laurels,
cinnamon, all growing in the Northern, and even
the Arctic regions. Then followed the glacial
epoch with its chilly exhibition of phenomena,
when even the South became Arctic, with Northern
bears and mammoths prowling about, and men hid-
ing themselves in holes. After that comes the
quaternary period, which brings us to our own
times, itself subdivided into two formations, the
diluvial or post-pliocene, and the alluvial or recent. The prehistoric men whom we have been
speaking of thus far, whether neolithic or palæolithic, are identified as diluvial, of quaternary times.
These geological distinctions we mention because
of this term " diluvial," as applied to the man whom
we have been speaking of as palæolithic or neolithic.
The other name for diluvial, that is, post-pliocene,
is so conceived as referring to the last portion of
the age which was previous to the quaternary, or
fourth age, and which is therefore called tertiary,
or the earlier, third age. This tertiary, like other

ages, is subdivided into various geological formations, of which the earliest is called eocene, and
the latest pliocene. Thus, then, we are to understand terms: when we speak of alluvial or recent,
and of diluvial or post-pliocene, we are in quaternary times; when we go back farther, crossing the
glacial epoch, we come to tertiary times, with its
various formations, the latest pliocene, the earliest
eocene. And if we hear, therefore, as we shall
soon, of a tertiary man, eocene or pliocene, we mean
one of whom traces are found in the corresponding
geological formations. While farther back still, if
an ancestor of ours existed in the ages of the secondary or primary formations, he would be called
by a corresponding designation. There is no question of such a being.

32. The results of all the archæology brought to
bear upon prehistoric humanity is to discredit the
idea that the human being then was of a
Results of Archæology. different species from the human being
now. The diluvial man's relative de-
gree of civilization marks no specific difference between him and ourselves. We might as well think
of classifying the Asiatic mountaineer of to-day
among things and men prehistoric. For the dwell-
ers on Mount Roraima are just now described as
persisting in the manufacture of stone implements;
at a time, too, when every possible advance in art
and industry is being made elsewhere, with the
help of steam, electricity and all manners of inven-

tion. Yet the said Asiatic is quite like the rest of us. Nor are we, by implication, very much in a state of barbarism, because we live contemporaneously with the stone age of Mount Roraima, 1888. And, in general, most nations have been found to use stone in the course of their history, the Israelites, Egyptians, Romans, the Indians, the Germans, the Lombards and Anglo-Saxons. Wherefore, the archæological ages of stone and metal seem to have been only relative, partial, local. Relative, too, and partial, is the antiquity which they indicate. And if there is any evolution in the question, it is only that of the sequence of stages in some nation's local development.

To despatch all the literature pertaining to diluvial man, we need add but little more. He has, indeed, played a conspicuous part in the hypothesis of evolution; and he still figures prominently in magazines and reviews for the entertainment of cultured classes, as also in the preliminary notions of children's text-books of history. We ought not, then, to close our obituary notice of him, without satisfying the reasonable curiosity of future generations. We shall just briefly look into the two remaining chapters of his record. And as, upon his withdrawal, his place was boldly aspired to by what is called the tertiary man, we shall say a word upon him also. The two chapters to which we refer are the palæontology and the anthropology of the diluvial or quaternary man.

33. Palæontology, or the science of extinct organic life, has shown us a series of animals which existed a long time ago; which were contemporaneous with man; and which have **Palæontology.**
now died out.   This seems to indicate a very remote antiquity for the prehistoric man who lived
with them.   Consider the long series of the cave-
bear, the cave-hyæna, the mammoth, the woolly
rhinoceros, the hippopotamus major, the Irish elk,
and such like beasts.   Their presence in prehistoric
man's time is betrayed by genuine fossils, or the
remains of their bony structures, which, whether
petrified or not, have been unearthed, or dug up,
that is to say, are "fossil."   And that man lived
with them is shown by his industrial remains, or
his own bones being found among theirs.   All
these species of animals are now extinct; and how
far away in the past must the man coeval with
them have lived and died !   " One's head is seized
with dizziness !" is the reflection of the modern
thinker, M. Renan: *On est pris de vertige !*

34. Scientists, however, criticizing this point a
little, have merely asked some pertinent questions:
How long does it take a species to die ?
Extinction of Whatever time it may take, did these
Species.
species wait for another, and gracefully
walk off the stage, each in its turn ?   Suppose the
environment did become unfit for them, and this,
indeed, was the chief cause of their extinction,
must it have been quite slow in becoming so ?   Or,

does all science admit that great convulsions took
place over the globe, great cataclysms which
changed abruptly all the features of localities and
of whole countries ? Or, let us with great liberality
suppose that the whole number of species now extinct, among which man lived then, amounted to
almost a hundred; and that they chose to die out
gracefully, one at a time; and that the species were
content with half a century each for its obsequies;
how many years would that require ? Less than
five thousand; not so far back as some chronologi-
cal tables put the birth of Noe.   As to their being
found petrified and preserved in soils, and rocks,
that does not prove the lapse of tens of thousands
of years.   It proves a little chemistry paying them
a tribute, which is due to their remains, no doubt,
but is not at all relevant to the question of their
antiquity.   Besides, scientists point out species that
die out, right under our eyes, and that too rapidly
enough; as the *didus* and *dinornis* of the islands
Bourbon and Mauritius.

35. The animals with which men lived have
served some observers as a guide to distinguish prehistoric times into three epochs.   First,
that of the great cave-bear; secondly, Subdivisions
of the Ages.
that of the mammoth; thirdly, that of
the reindeer.   This succession, however, being
ill-substantiated, gave way to an archæological
classification, taken from the stations in which industrial remains were found.   Four periods have

thus been named respectively from St. Acheul,
Moustiers, Solutré, La Madeleine.

36. Anthropology, strictly so called, considers the
prehistoric man himself; and, finding in certain
Anthropol-   fossil bones and skulls a type of man-
ogy.   kind very different, as it appears, from
the normal type of the present, it has supplied the
evolutionary theory with an important link in the
question of our descent. Various signs are noted
in those skulls, indicative of an inferiority to the
man of our time, physically as well, no doubt, as
intellectually. Those found in the caverns of
Engis and Neanderthal have become famous, if
only for the number of scientific monographs written upon them, to show forth the low state of humanity exhibited in their conformation. For instance, they are dolichocephalous, that is to say,
long-headed; the longitudinal diameter being ab-
normally longer than the transverse diameter.
Many other points besides this are brought to bear,
anthropologically, on the question of our descent;
which is so illumined, in consequence, that, not to
mention others, Max Bartels has brought together,
in a monograph of nearly one hundred pages, the
literature and notices of men with tails. We must
confess that we have not made any closer acquaintance with this valuable work than to read the bib-
liographical record of it in the Smithsonian report
for 1885. But that does not dispense us from pay-

ing a due regard and close attention to the points which anthropology has noted in these skulls.

37. The first observation that occurs is this. No sooner was dolichocephalism, or long-headedness, noted, than a comparison was instituted at once, in the interests of science, with the actual races of mankind, of which we shall treat expressly in the second part. And it was found that existing men show every type and measurement, as well of this cranial conformation, as of its opposite, brachycephalism, and of every other. M. de Quatrefages transcribes long lists of measurements which show this. In the second place, a number of other anatomical elements, thought to be peculiar in these fossil skulls, such as the superciliary prominences, the small and receding forehead, the form of the ciliary arcs, the amplitude of the occiput, are found to be but the individual and accidental varieties of men living among us. Neither the low-minded amongst us, nor the high-minded, nor even distinctive eminence in cultivation and genius has appropriated any exclusive form of cranium. The form can be modified before birth, and the peculiarities become congenital. It can be modified after death, and they are posthumous; physical and chemical agents so far affecting the skeleton as to change the proportions. Other causes operate during life ; and they are either artificial in their nature, as the forced compression of the skull, a practice still holding among certain

(marginal note: Physical Varieties.)

tribes; or they are natural, as heat, light, actinism,
moisture, atmospheric contamination, drink, food,
resources, scenery, degree of natural security, con-
sanguineous marriages, sickness, and the like.

It would appear that a great many elements were
necessary to conclude a logical argument here. In
the absence of the argument, what becomes of our
poor savage species, the ape-man, who yet must
be found somewhere, if evolution is to hold its
ground ?

38. There is one resource left. If the diluvial man
of quaternary times is nowhere at the service of
Tertiary Man. evolution, perhaps a tertiary man of
the times gone before would be so, if
only he could be found. To the satisfaction of a
goodly number of scientists, French, German and
others, such a prehistoric being of the tertiary age,
both pliocene and eocene, has been found, and that
several times over. Not that he himself has quite
shown himself. His friends admit that. But he is
hypothetical in other things, which certainly have
been found. To find himself then is only a question of time, when a future day will reveal him;
and faith in the vindication of science is long-suf-
fering enough to await that day in patience. The
things, in which the tertiary man has betrayed himself, are flint-chips, and flints burnt, and irregular
incisions made in the bones of animals, all of which
are found in tertiary formations, and belong to tertiary times, and therefore—reveal a tertiary man.

39. It is a little singular, on the face of it, that his own bones do not appear just as readily as theirs. There is no natural law requiring the more rapid consumption of human bones than of beasts' bones. If one kind are fossilized, why not the other? Cuvier demonstrated that the bones of ancient warriors show no more readiness to decompose than those of their horses.

40. Still, not to be wanting to the true spirit of scientific thought, let us contemplate those flints or stones of the tertiary age, some with what is called a conchoidal fracture or a bulb of percussion in them, such as would result from an intentional blow, and therefore indicating a person who intended the blow; some apparently scorched, as having their outer surface disintegrated; and therefore indicating a person who scorched them. Now, we are to consider these signs as being so indifferent in their character, that they point indeed to a man who made them, but they postulate only an ape-man, an *anthropithèque*, one who knew just enough to do that, but knew no more and knew no better. This is the logic which satisfied the French scientists in the gathering at Grenoble; and they agreed by vote that the existence of a tertiary being was now proved.

41. We cannot help thinking that other scientists, of quite an opposite school, have some ground to be well pleased with this course of reasoning. The form of the logic used impresses the mind favor-

ably. We fancy that we see in it a strong reassur-
ance of some general revival in sound
Form of the
Logic.   logic and solid thought.   For if, from a
chance percussion in a chance stone,
such a fracture as appears over and over again on
our roadways, made by the hoofs of horses, or by
the rolling of wheels, the sagacious minds of men
can discern the presence of an unknown being
whom no other sign manifests, and can just measure
his intellectual capacity and portray his physical
build, it is quite evident that many of the noblest
sciences are in a fair way to being reinstated. And,
as for teleology in particular, as for theism and
theology generally, the good time is coming when
no man's mind will fail to see, in the marks of
beautiful order impressed on the world and the universe, a magnificent testimony to the existence of
One who must have intended it, and made it, and
almost a description of Him, who having made it,
is now preserving and governing it.   So much for
the form of the argument, or the manner of the
logic.

42. Now a word on its matter.   It has been asked,
in a somewhat critical spirit, whether similarly
broken stones, which are found to be
Matter of the
Logic.   scattered about on a shingly beach,
argue the presence of men all about
there to do the breaking?   Again, M. Arcelin, a
French scientist, most prehistoric in his tastes and
specialties, picks up, in the argillaceous silex of

the Maçonnais, flints of precisely this description, with the fracture which is thought to reveal an intentional act of breaking, and yet which is referable in this case to atmospheric agents. Besides, were there no hoofs, no tramping, no rolling, no crashing, in the days of the great mammalians, and among the gigantic disturbances of past ages; when in the ordinary flow of those mighty volumes of water, that eroded the primitive beds of rivers, the collapsing of huge blocks of silex brought about collisions, more than are needed for myriads of conchoidal fractures and bulbs of percussion to be laid out on the bottom of the waters? Again, if any tertiary man broke some of the flints, he must' have lived at the bottom of the sea to do it; for those exhibited by M. Cels, just recently, to the Anthropological Society of Brussels, by way of proving the tertiary man's existence, are taken from lower eocene sediments which are observed to be altogether marine, containing mollusks, fishes, chelonians, and the like.

43. Nor does the disintegration of the surface in a flint seem to be due to fire alone. If it is, however, were there no prairie fires, no forest fires, breaking out spontaneously then, as they do now ?

44. And, again, if we inspect those fossil bones of animals, with the irregular incisions made in them, are we inclined to believe that a man only could make an incision, particularly an irregular one ? It may be that wild beasts preyed upon one

another then, as they do somewhat freely now, and
that with teeth which could scrape one another's
bones pretty incisively.   Scientific men go to the
trouble of pointing out effective teeth of that age,
such as seem to suit the incisions exactly, those of
the lusty beasts called *carcharodon megalodon, sargus
serratus*, and others.   These dreadful names insinu-
ate nothing but teeth !   In fine, the critics urge the
importunate question: Did the beasts of prey spare
man himself, and not scrape his bones for him ?
If so, where are they,—be they scraped or un-
scraped ?

45. It is easy to ask questions, and for irreverent
minds to ask irrevent ones.   But professors of Berlin, and scientists of the French school
itself, nimble as that school is in its logic
and its fancy, have thought it was rather
easy to make random statements, and scandalize
science by settling things with a vote.   If things
are true, they need no vote.   And the only result
of the pronounced discredit which this controversy
has thrown upon the tertiary unknown, is to show
him unknowable, probably because he is not there.
Twenty years of contention about him have left
him where so many are leaving the missing link
generally; and that is nowhere.

Voting in Science.

46. It really makes very little difference where the
first appearance of man is placed, and how it came
about, if only he was there.   We shall learn much
that is useful, when we ascertain where it was, in the

order of geological formation, that he did first appear on this globe. He will throw as much light on geology and the other sciences as they throw on him. At *Man and Geology.* present, he is not shown to have walked this earth at any point farther back than the diluvial period; as M. d'Estienne just now affirms,"there is no geologist of note who admits any longer even the possibility of man having existed in the lower tertiary age." General considerations forbid us to expect that we shall ever find it shown. For, if man is the head and completion of the physical and organic world, as all admit, and evolutionists no less than others, he could not appear till the physical conditions of things, and both the vegetable and animal kingdoms, had received their just development. As competent science affirms, he must be the last in both the stratigraphical and the palæontological lines. Before that, he were an anachronism.

47. So, to conclude this criticism of the prehistoric ape-man, whose geography and chronology we have subjected to a little analysis, we may express ourselves in the fine gen- *Man and the Universe.* eralization of Agassiz. He says that, as the reptiles of the secondary age are in no respect descended from the fishes of the primary or palæozoic age, so man in the fourth or quaternary nowise descends from the animals before him in the third or tertiary period. The link by which they are all connected is of a higher and immaterial nature.

Their connection is to be sought for in the view of the Creator Himself, whose aim in forming the earth, in allowing it to undergo the successive changes which geology points out, and in creating successively all the different types of animals which have passed away, was to introduce Man on the surface of the globe. Man is the end towards which all the animal creation has tended from the first appearance of the palæozoic fishes.

We have finished with prehistoric humanity. We shall review next, in a more constructive spirit, the anthropology of the actual races which are; and shall find in them the key to all the prehistoric difficulty which has been.

ANTHROPOLOGY.—*Continued.*

CHAPTER II.

ACTUAL RACES IN HISTORY.

48. We have considered whether, in the past, there ever existed a species of men, different from that which we know of now. This was the question of prehistoric anthropology. It still remains to be seen whether, in historic times, any man of a different species from our own has existed, and can exhibit in his person the link which is sought for to connect us with a lower order of animals. The most imperfect races of mankind are judged to be those in Australia, and others, such as the Bushmen, in Africa. But these are now pronounced, by the most unexceptionable science of the day, to be altogether men of our own organization. So that, if we go by the authority of scientific men, the question is closed. There is not, and there has not been, any species of mankind distinct from the one which we know. All men are, and have been, of one formation, one organization, whether they are looked at anatomically, physiologically, or intellectually.

47

49. But if, instead of merely taking the authority of scientific men, we examine the scientific results for ourselves, we shall derive profit in two ways. On the one hand, we can enjoy the advantage of seeing the facts for ourselves, and of concluding that there is no color whatever, in the observations of science, for the hypothesis of an ape-man. On the other hand, a philosophical view opens out before us regarding the course of man's progress upon earth. We are thrown back into the same vein of thought with which we started, that evolution of man's history, from its origin onwards, through its varied and divergent course. We contemplate down many an avenue and vista of human history thus distributed over the face of the globe, and down through the ages of time, how the whole progress, so divergent to begin with, is converging towards a final reunion of the human family, when God's designs shall have received their entire accomplishment over the children of men. This is a physical side to man's ethical and intellectual history. It is credited to the science of anthropology.

50. Let us observe, then, that in the collection of individuals, called mankind, there are many differences, as well anatomical and physiological, as intellectual and moral. Organs and functions, ways of thinking and acting, are all found to be diversified in various natural groups, which are called Races. Now, what do we affirm? That, in

spite of all differences, the races are of one Species. And, moreover, being of one species, they are inferred to have had one common origin; which, in biological matters, means that they' sprang from one primitive pair. If this is so, scientific evidence corroborates, with its process of induction, the documentary evidence presented in the narrative of Moses. We begin by recurring to biology (No. 104, etc., below) for the explanation of these terms: species, race. Then we shall apply them to the subject of anthropology.

51. By the term, species, we mean a collection of organic individuals more or less resembling one another, in their external aspect or internal structure; productive in their **The Term, Species.** unions among themselves, so that they perpetuate the same collection in nature, by generating other individuals of their own kind; and one of the consequences thereof is, that originally all can have descended from one primitive pair, identical in kind with them. This description of species, which is evidently founded in nature, and is exemplified in the whole of biology, is not to be confounded with another use of the term, species, whereby it is taken to signify any mere class. Thus, a distinguished palæontologist, attached to the U. S. Geological Service, uses the term, as if in biology we signified by it any mere group. We classify, he says, organic beings, as we would classify bottles; and therefore, he concludes, there is no reason why one

species should not turn into another, by what is called " descent," or "transformation of species;" just as among bottles we can reassort classes, and have in one group to-day the bottles which we had in a different group yesterday.   Here we must re-mark that the term, species, is taken in quite a different meaning from what it must have if we are to discuss at all the question of the descent of species.   And indeed any one arguing so, on this subject, commits the logical error which is called equivocation, that is, playing on the same word in two different senses.   Considering the scientific and philosophical gravity of this error or sophism in particular, we should desire nothing more than to see it first pilloried, and then petrified in every text-book of grammar, rhetoric and logic to the end of time; till a future age of anthropologists shall unearth it, and wonder wisely, what kind of pre-historic barbarians devised such a fossil piece of industry, and made it, and used it.

52. In the scientific idea of species, it is not any resemblance that determines the class.   The like-
The Test of ness among individuals may be more or
Species.   less.   It may be lost so far in apparent unlikenesses, that other beings of a different species may come closer to the type, in appearance, than organisms of really the same species; as in all classes of things we see that extremes touch, or even overlap one another.   In biology, it is a like-ness indeed that determines the species; but it is a

radical and primary likeness, one so deep as to be tested by nothing less than a deep-seated physiological or vital function, which of course is a radical quality. All physiological qualities are deeper than morphological proportions, or anatomical structure; these latter do not exist but for the former; external proportions and structure and the organs of life do not exist, but for the vital or physiological functions to be performed through them. If there is any precedence between physiological function and organic structure, it is not the organ that is prior to the function, but the function that is prior to the structure, and is the reason for its existence.

53. Now, there is a function of reproduction, whereby a living organism reproduces its kind. This is a law in every order of living things: " Like produces like," *simile generat simile.* If that class only of living beings is called a species, which can unite and reproduce its kind, you see a physiological quality is referred to, very different from the external likeness among bottles, which Professor Cope offers to define species by; or the structural likeness between man and the ape, whereby other Professors suggest that species should be determined. The power of reproducing its kind, or generative productiveness, is the test of species.

54. In biology, they would illustrate the matter thus. The animal class called the horse has prop-

agated itself from time immemorial. So has the
ass. The mule never yet—perhaps to our great
relief. Horses are a species. So are asses. But
the hybrid mule, which is a cross-breed between
the horse and the ass, is not a species. Hence we
see that organisms can exist without constituting a
species of their own. Others exist, and do make a
perpetual family of their own. While no families
ever pass over from one line of propagation to
another, giving us what is called a descent, or transformation of species. This we shall see in biology.

55. The idea of race is much easier to apprehend.
It originates in the fact that every species admits
of varying traits in the individual, as dis-
The Term, Race. tinguished from any other of the same
species. Indeed, no two individuals are
in all respects alike. The specific likeness, remaining
common to all, is modified accidentally. Now, these
accidental modifications may keep within certain
normal limits, usual in the species; or they may be
exaggerated or diminished, moving in a positive or
negative direction, outside of a usual area of un-
dulation, observed in that species. So doing, they
become exceptional. And the individual which
bears an exceptional character of this kind is
called a Variety. Should this variety transmit its
peculiar modifications to other individuals, by way
of descent, there results a line of posterity marked
with an hereditary divergence from the common
type. A posterity like this constitutes a race;

which is thus seen to be within the species, to be perpetuated by generation, and to have taken its rise in an individual variety. Meanwhile, the fertility or productiveness, which is radical in the nature of the entire species, remains uninterrupted among all the individuals, whether of one race or another.

Let us give some instances of all this from general biology, whether of the vegetative or the animal kingdom. The sketch of what we observe there will serve for the argument of analogy, by which we shall draw conclusions with respect to man. Analogies of the Lower Orders.

56. Plants of the same species vary in many ways. The organic elements are seen to be differently associated and combined. Acids may diminish or disappear, and be replaced by sugar, with a sweet taste and perfume; and these developing will characterize distinct races of vegetables and fruits. The plum, the peach, the grape, are instanced as having been subjected to cultivation, improved by means of the modifications superinduced, and then perpetuated as the agreeable fruits which we know them to be.

57. The vital functions, too, of a vegetable species may become remarkably altered. Thus, in different races of grain, the power and rapidity of growth vary as widely as one to three, some taking three times as long as others to grow, while all are of the same species. In temperate climates, barley

requires five months to germinate, to grow and
ripen, whereas, in the cold of Finland and Lapland,
it accomplishes the same phases of growth in two
months. The power of reproduction, also, can
vary so much that some roses will bloom two or
three times a year; and there are strawberries that
keep in fruit nearly the whole year.

58. To be quite precise in this matter, we may
sum up with Wigand all the known variations or
modifications under the following heads.
Variations. There are chemical alterations, as in the
color, acid, sugar, ethereal oil, etc.; anatomical
alterations, as in the covering with hair, texture,
thickening of cellular walls, etc.; the physiological
function of growth, or enlargement of the entire
plant, as well as of each single part, without prejudice to the essential proportions of form; also the
periodical functions of the producing of foliage,
the time of blooming, of ripening, of dying; morphological variations, in the direction and relative
length of the shoots, in the form of leaves, or in
the indenting of their edges, in the relative dis-
tances of parts, in the number of flowers, etc.;
modifications, in short, more numerous than we
care to exemplify. Now, all these changes are
actuated, perpetuated, or reduced again, whether by
the mere operation of nature, which has been
called "natural selection," or by the sagacious de-
signing of experimenters, which is much more
effective, and is rightly termed " artificial selection."

Yet no alteration is discernible in the specific character mentioned above (No. 52, 53). That exhibits an identity, unity, harmony, amid all varieties and races, howsoever far they wander from the normal, middle type, howsoever much they fluctuate, or undulate above or below the mean level.

I would gladly go through a similar series of instances in the animal kingdom; and this would be the more interesting, as the analogy which they supply is closer to the human species, reaching as they do into the order of sensitive instincts. But space and time forbid it here. So let us pass on to the argument of analogy derived from these facts.

59. The argument, which is called analogy, may be taken in a loose or in a strict sense. In the looser meaning we have it constantly referred to as "analogies of nature," supplying funds inexhaustible for the similes **Argument of Analogy.** of the poet, the moral lessons of the philosopher, and even sometimes for the embellishment of a scientist's theory, when he grows oblivious of his exalted responsibilities. It is only in a strict sense that science has anything to do with the argument of analogy; which means that, given the identical data, under the same bearings in two otherwise different classes of being, these identical data so considered may be taken as premises to draw conclusions, which then apply equally to both. Thus, given identical conditions in the optical organs of a

bird and of a man, the conclusion to be drawn with regard to some object seen will be the same in both. Granting an ape to have lungs and a man to have lungs, then in similar conditions each is liable to grow consumptive. Considering, on the one hand, the perceptions or instincts which are developed in sensitive organs, and, on the other, that kind of knowledge which is gathered by an inorganic intellect, the generic idea of perception or knowledge will apply to both, to the sensitive instincts, though they be only in a fowl or à pointer dog, and to the thoughts and reasonings of a man's mind; while the perceptions in the two cases remain specifically different. In short, to illustrate what is meant by analogy, let us take a familiar, every-day instance. There are ten cents in a dime, and ten cents in a dollar, but differently. The equivalence is exact in both. In the dollar, however, there is something more; and the ten cents that are in it are not there as in the dime. Yet, what they are worth under the one form, they are worth under the other. This is an identity of analogy in two different coins, the dime and the dollar. What it means and comes to in different organisms, as the ape and the man, we shall see under the head of biology (No. 197); as well as the surprising conclusions that have been drawn from it; as if two beings that are identical under any aspect of analogy must have come one from the other directly, or both collaterally from a third. I refer to the theory of what is called the descent of species. It is chiefly built up on this argument of analogy.

60. Now, applying this style of argument to our present subject, we can proceed thus : If, in the species of organisms, whether plant or animal, there can be so many differences among races, without prejudice to the identity of the species under which they are ranged, there can be as great and as many differences among the races of mankind, without any injury to the notion of an identical species containing all. In point of fact, we shall now find the differences among human races to be much less than those noted among animal and other organisms. The application, then, of the indirect argument of analogy, brings us face to face with the results of observation; and these furnish the direct argument regarding the human family.

61. Thus we observe that, in stature, the extreme variation among men is from that of the Patagonian, who averages nearly six feet high, to that Results of of the Bushman, only four and a half, Direct Obser- or, as the scientific journals lately report, vation. that of the Akka tribe, who are apparently but four feet in height. These varieties, which are the extreme ones, are to one another as two to three, represented by the ratio two-thirds. Now, on the other hand, in the animal kingdom, the variation in stature is found to be as one to five between the small spaniel and the great St. Bernard, the former being only one-fifth of the latter. Yet that does

not prejudice the identity of their canine species, in the strict scientific sense of that term, as explained before (No. 52, 53).—I take it, you understand quite well how this comparison between man and the brute creation comes to be made so constantly; and I need not apologize for instituting a scientific comparison between them. It is because of the analogy that exists between them, since all are in the one genus of living organisms having animal life. But, as to their respective species, man is specifically a rational being; whereas the rest are irrational, or brute.

62. It is to be observed that these extreme varieties in stature among men are not separated by any wide chasm between, as is the case in almost every respect between man and the ape; where no scientific trace of a bridge between, which still connects them, or of a common point behind whence they could have diverged—no "link," as it is termed —has ever been detected. Here every size between the extremes in mankind is represented by existing races, all thus merging into one another insensibly. The same holds with regard to colors. There are extreme colors among men, as there are in species of the animal kingdom. But every shade between, of black, white, yellow, red, is also exemplified. The blending and mixing of hues and tints is equally noteworthy among men and among irrational beings of any identical species. Dogs and horses, varying as they do from one extreme of

color to another, assume among other things a white hair on a black skin. Among poultry, the domestic fowl of French breed has a white skin; that of Cochin China approaches yellow; and there are black fowls with a black skin, while the silk hen of Japan has a dark skin under white feathers. Now, with human beings the accommodations have been just as striking and as numerous. There are white races which are black; "all black men," it is said, "are not negroes." That is to say, the Hindoo is of Aryan race; the Bisharee and the Moor are of Semitic blood, all therefore of a white stock; yet they have assumed the same hue as the true negro, and even a darker hue. In fact, on any white person's skin, the spots called freckles are said to present the same characteristics as the negro skin.

63. These matters, and many others which show race merging imperceptibly into race, and therefore without any specific chasm between; which take place right under our eyes, and therefore are of immediate scientific evidence, lead us to the following conclusion. Physical and Physiological Conclusions. It is clear, by the genuine law of inference, or induction, that, in similar conditions holding in past times, similar changes must have been taking place as take place now; and some varieties thus springing up must have been perpetuated in races. Here the argument of analogy coming in emphasizes our conclusion with regard to man, by showing that the most divergent human

races are, after all, less different than races of a given species in the lower kingdoms, vegetative or animal. Hence, the inference stands confirmed that all humanity has been of one species from the first, and could have descended from one primitive pair. This proves the identity of the human species from the physical side.

64. But the test above all is that direct one, a physiological fact, which proves identity of species everywhere,—the fertility of the human races, their power of blending all over the world. Distinct species refuse to blend; human races do not. Nor do the latter deteriorate by blending, and die off in sterility, as hybrids do (No. 133 below): on the contrary, they improve As a species, then, the races of mankind agree in all the anatomical and physiological characteristics which go to make one family.

65. We should not omit their intellectual qualities. Here the same unity is manifested in many lines of demonstration. There is language, which is necessarily and primarily a vehicle of thought from mind to mind—thought in its entire range of objects, abstract and spiritual, as well as concrete and material. This language or speech is a power in which philology is but beginning to reveal the beauties and hidden depths of intelligence concealed under its forms,—a power which reflects as a mirror the thoughts of our inmost soul right into the soul of our fellow-creature ; singing in joy,

Intellectual Qualities: Speech.

mourning in grief; inscribed on monuments in desert isles, printed in the page of civilized peace; bearing down through all time and space upon earth the tenderest thoughts and feelings from the hearts that are now no more.

66. To be quite candid, we should not disguise the fact that a certain species of modern science thinks very differently of this noble faculty. It has referred the origin of language to a prehistoric ape, which tried to take off the savage howls of other sav- Talking Apes and Sub- merged Con- tinents. age beasts; and improving by the practice became the *anthropithèque*, whose acquaintance we made before (No. 40); and in that capacity gasped for articulation, and got it; becoming therewith an Aryan ancestor of ours, who pitched his tent in Asia. But as no Aryan *anthropithèque* has now been found in Asia, it has been thought safer to pitch his tent for him in a submerged continent, calling it Atlantis. To quote a Mr. Heath, F. A. S. L., who dates from some time and portion of creation which at present we forget—Mr. Southall reports him: " It is known that there were anthropoid apes; it is knowable that they gasped after articulation, and that those who attained to it are Aryans, whether of Asia, or of the submerged continent of Atlantis." This gentleman is doubtless a little trenchant in his style; making up with the pride of assertion for the poverty of fact. Other scientists, equally poor, are less proud. Even Professor Haeckel is more meek amid

the poverty-stricken archives of prehistoric evolution. And as to Mr. Darwin, he is inimitable in his unfailing humility and sweet *naïveté*, when, alluding to the unhappy poverty of facts to favor his doctrine, he says that doubtless the proofs are buried deep beneath the waves, in continents submerged, and so forth; and he continues: " This manner of treating the question diminishes the difficulties considerably, if it does not cause them to disappear entirely."

67. Gentlest suavity of guileless humanity! Ah! that bosom of the deep Atlantic, how dear do even those chilly depths become, if only the beloved object may still be there! How many treasures,—to quote the tragic poet —besides the son of York, are in the bosom of the ocean buried! Wrecked, not on the Goodwin sands, but on the sands of time, of hypothetical times and times, a continent lies there, and an hypothesis, nestling in the heart of a fossil continent! You have heard the pathetic episode. Now listen to the tragic palinode. The expedition of the " Challenger," sent out by the British government, publishes in its late reports that no such continent as an Atlantis ever existed! Is that so? Really, is a watery grave refused to the man-ape and his fugitive hypothesis, when they begged for nothing more than the cold hospitality of a five-mile depth in the ocean? It is even so! Mr. John Murray, speaking with all scientific authority upon this oceanic ques-

Mr. Darwin et alii.

tion, says: " He is a bold man who still argues that in tertiary times there was a large area of continental land in the Pacific, that there was once a Lemuria in the Indian Ocean, or a continental Atlantis in the Atlantic!" Still, who knows? A Mr. Heath may still contend that they are "knowable!"—a euphemism, you know, for the unknown, and probably unknowable. Can you or I dispute such a knowable, especially when so desirable? As well argue against a fond wish by the rule of three, shake hands with a ghost, or knock down a phantom! 'Tis " a great lesson!" exclaims the Duke of Argyll, touching off the whole comedy in an interesting review which he entitles so, " A great lesson!"

68. To proceed, if speech be taken in another sense, as a certain collection of sounds or words, called a special tongue, or language, it **Tongues.** may be thought that all tongues should be traceable to one original tongue, or stock, if all races came from one orginal pair. In point of fact, philologists are now placing side by side, and are showing to be strictly akin, various tongues which before were thought irreducible in grammar, as well as vocabulary. The degrees of kindred are already marked in the Aryan family of languages. It remains to be seen whether all the families, monosyllabic, agglutinative and inflexional, will yet be brought to a common centre. But if they cannot, that will only go to prove another point of historic fact, an alternative thesis equally worthy of scien-

tific verification, a fact which one record distinctly
testifies to, and which many other lines of tradi-
tional and documentary evidence are converging
to; that once there occurred a violent breaking-up
of the human family, a catastrophe in the social
order, well known to us in the Mosaic narrative as
Babel.

69. Besides language, and other manifestations
of intellect, there are identical moral characters
which are wanting in no human race.
Conscience. Conscience is there, with its rewards
and its remorse, its hopes and its fears, the same
that have inspired the white-robed army of heroes
and martyrs, the heroes in thought and action, the
martyrs to honor or to truth.

70. Here, too, we have been rudely awakened to
the fact that the moral problem contained in the
beautiful sentiments of conscience has been solved
by that intrusive species of modern science, as with
the touch of a master.   An eminent observer points
to his dog, in the act of refraining from eating its
master's dinner, when it might.   See, says the master, there is the moral sentiment, in its first stage of
evolution—conscience in embryo!   Such is the explanation vouchsafed us for the moral order and its
origin; and it is an explanation, we must confess,
which suggests pregnant reflections.   They become more fertile still, upon the further statement
being contributed to the question, that the dog looks
up to its master as its god!   A compliment, indeed!

though not so disinterested as you might think, if the master meanwhile is looking back at the brute as his own progenitor.

71. This explanation notwithstanding, the moral manifestations remain distinctly and obviously human. Charity, for instance, is there; and that which is called philanthropy. They have animated loved and loving souls, whose very names send a thrill through the generous heart. Nor could we undertake barely to enumerate the great divisions of those armies which have graced the field of charity, truly a "field of the cloth of gold" in the wide domain of our history upon earth.

Charity.

72. It grows a little tedious to be told again by this irrepressible science that even charity, the ornament of our humanity's bosom, is readily explained by—what we should like to omit, but what the symmetry of solemn repudiation on the part of sound science will have us record. Well, this importunate science—this infant terrible and uncontrolled, sprung from the cultured thought of a kindly man—explains charity by pointing to dogs licking one another, oxen similarly, monkeys ditto!

73. Finally, there is the religious idea, everywhere leavening the races and exalting the nations. It is the same which, in educational systems of the day, is rather freely relegated into the realms of poetry. Yet, even so, it is indeed the truest poesy, raising the mind to an Infi-

Religion.

nite, Necessary and Invisible One, throwing nations down in prostrate adoration before a hidden
majesty, and inspiring with loftiest speculations the
mind of a Plato and a St. Thomas Aquinas.

74. Out of justice to it, and to all parties, we will
mention the last conceit which we condescend to
stigmatize in this species of science, unutterable
and indescribable. To explain religion, what does
it exhibit? A dog. To escape the divine, it hugs
the brute. There! look at that dog, quaking at an
open parasol, which the wind is just sensibly agitat-
ing, as it lies upon the lawn. The dog sees the
parasol moving: it does not happen to feel though
that it is the wind which causes the motion. Between ourselves, why should it, for the beast is
only a brute? So the intelligent creature begins to
quake and yelp at the phenomenon, as preternatural;
because, says Mr. Darwin, its cause is unknown.
Here, says he, here is religion for you—trembling
and quaking before the unknown—the preternatural
—religion in germ!—the embryo of all that high
and sublime knowledge of God, which invites wor-
ship and adoration, yes, and fear and holy sacrifice.
All this is in an agitated umbrella, and an agitated
dog yelping about it!

75. We feel compelled to pay a tribute of respect
to the generation of science which is now in pos-
session of the field, and observe of many of our
scientific men that the Darwinism, which they still
promulgate, is not that of Mr. Darwin. His essay

did indeed bring in a tribe of systems; and he set the example of "wrapping theories up in facts," which is not, as Wigand observes, quite the same thing as proving them. But we beg to leave it on record, that some of the systems he helped to bring in are sufficiently respectable compared with their pioneer. His attempt was only a pioneer of theirs, not an ancestor; connected with them by analogy, not by descent. This we can say, if you like as a compliment, but also in truth; in every case, however, by way of conclusion and taking leave of this unutterable school, for the present. I hope it is clearly enough established that all the races of man are specifically identical in moral and intellectual characters, in physiological and anatomical qualities.

76. Only one thing remains, and that is to show the process whereby the races could have come to be differentiated so much, Differentiation of Races. if they ever had one and the same origin; and differentiated too within the space probably of five thousand years. Are any reasons assignable for considering that they originated in mere modifications of the same stock; so that the modifications, being perpetuated by generation and becoming fixed by inheritance, have given us all the present races as a remote posterity of the same primitive pair? The variations in question, which distinguish the races, are such as stature, figure, color, shape of the cranium and of the face, weight

of the brain, rapidity of development, duration of life.

77. The answer is simple and positive. The various conditions of life in which men **Conditions of Life.** have been placed are amply sufficient to account for all the modifications that characterize human races. By conditions of life we mean the soil, cold, heat, dryness, humidity, light, drink, food, resources and the like. Such various conditions in the environment have made acclimatization necessary, and naturalization possible. That is to say, new conditions of life have appealed to a certain self-adjusting power in the human constitution, and have brought about in it a degree of harmony between the constitution and its environment. This approximating harmony is called acclimatization. If it is rendered perfect, it is naturalization, and implies that our common human nature has put on in the given circumstances some special characters and aptitudes; which make up an acquired or second nature, exactly adapted to the new conditions of life.

78. Let me illustrate these adaptations by recurring for a moment to our old argument of analogy (No. 59). Sheep have a woolly fleece, as we know to our great comfort in winter clothing. But in the heat of Africa the sheep change their fashion, and put on a short and smooth hair. Just the reverse takes place with the wild boars, when they pass from a warmer to a colder climate. They are accustomed

to wear a hairy coat: but on moving up into the high plains of the Andes, and the rarefied and chilly atmosphere there, they change their mind, and take on a kind of coarse wool. Or, to illustrate the same process of adaptation in man himself, both naturally and artificially, I will cite a note of M. Flinders Petrie, which you may take for what it is worth. It is amusing, but suggestive: "We all know how translucent flesh is to strong light, and it can hardly be doubted that the rays of the tropical sun would light up a white man's inside considerably; whereas black skin would stop out the solar energy of light, heat and chemical rays effectually. Skin heat is of no importance, as perspiration can always keep that down. May not the oiling of the skin in hot countries be partly to make it reflective, so that it should absorb less heat? And may not the regard that white races have for clothing be partly for the purpose of keeping the insides of their bodies sufficiently in the dark?"

79. To return to our question, I say that an acquired or second nature will be the result of all the combined influences, which act in a given environment upon our radical human nature. Ordinarily speaking, too, only one particular race will result as the final outcome of perfect naturalization in given conditions of existence.

80. Here, then, the character of race is seen to be made up of two elements, the fundamental or radical nature common to all, and the acquired or

second nature of acclimatization or naturalization. The proverb may be referred to here: " Custom is a second nature." But the custom we **Radical** speak of now is physical, physiological. **Nature and** It reacts upon the structure and form, **Racial Nature.** affecting the anatomy and morphology. It explains, also, much of the special pathology which distinguishes a race; that is to say, its liability to disease. For suppose that a disturbing action which is the cause of disease begins to work upon a system. It may affect the common, radical element in our human species; and then the same disturbing cause will produce radically similar effects upon all human beings, notwithstanding their racial differences. This we experience in certain epidemics that spare no races. But if this disturbing action is spent upon the acquired and special element in a race, the same cause will not produce the same effects in different races, and the disease will be more or less peculiar to certain groups. Thus the European is singularly susceptible of marsh fevers, to which, on the other hand, the constitution of the negro is very indifferent. But, in moist atmospheres, the negro is observed to be very liable to consumption, while the European enjoys a comparative immunity.

81. Obviously there are some variations which are quite normal in our climatic condi- **Vitiated Conditions of Life.** tions, as certain varying degrees of moisture or dryness, of heat or cold. But

there are deviations, too, which must be considered abnormal. In other words, there is a limit, beyond which the great pliability of our constitution fails to adjust itself. The environment is then said to be vitiated, whether naturally or artificially. The great estuary of the Gaboon is naturally so insalubrious that it is destructive to even the native negro. Accidentally, the marshes of Corsica, the Maremma, the Campagna di Roma, are to be ranged in the same class; but they are within reach of improvement. The hearts of many great cities, like London, Paris, New York, are vitiated artificially. It is reported that a Parisian of unmixed Parisian blood cannot be traced farther back than three generations. The city families are extinguished in that short space of time. Similarly, at Besançon, families die out in less than a century; and they are insensibly replaced by healthy ones from the country.

82. Amid the immense variety of conditions, to which human life has been exposed, our species, located at first within a very limited area, has now been distributed over the entire globe, and has adjusted itself to every shade of difference, from either polar region to the equatorial zone. Let me call your attention to its classification into hunters, shepherds and tillers of the soil. The hunters are they who first spread in every direction. We have seen them in the Indians of this country. Their mode of life,

Hunters,
Shepherds,
Farmers.

the necessities which it imposed, as that of obtaining food by the chase, the instincts which it developed of wandering on and on still further, carried the hunters far away, oftentimes into desperate conditions of existence, both natural and social. This it is which the palæolithic and neolithic relics give us to understand; and in a manner pathetically portray to our eyes. In migrations by land, hunters or warriors were ever the vanguard. Seakings led the way by sea. And the vanguard had to pay, with their own personal suffering and loss, the price of that acclimatization which their posterity might survive to enjoy. After the hordes of hunters or warriors, it is plausibly suggested that the life of nomadic shepherds would be the first step towards settling down; and they would thus begin to agree in their manner of life with those who had never migrated. Finally, with increase of population and a greater degree of security, the shepherd would settle down still more, and till the ground, opening up its resources as an agriculturist or farmer.

83. What this acclimatization might cost the pioneers, we may estimate by the analogy of the lower organic kingdoms. Spring wheat has been made to change its times and conditions into an autumn wheat; but the effort entailed the loss of nearly three harvests. To make wheat grow at all in Sierra Leone cost many more. Certain European fowls,

Cost of Acclimatization.

familiar to the English Christmas table, were introduced into Bogota, South America, and they got
acclimatized there only after twenty years of effort;
so that practically twenty generations were lost,
before the little that survived each time had sufficiently adapted itself to increase finally and pros-
per.

84. Using the force of this analogy, we need not
be surprised at the condition of the settlers in
Guadeloupe and Martinique. These islands are
the most fatal of the West Indies. After ten generations from their original settlement by Euro-
peans, they are still dangerous to the inhabitants.
However the population now manages to increase
at the annual rate of one-half per cent. Algeria
has been but recently colonized, and it is still fatal
to Englishmen, Belgians, Germans; but it is becoming tolerable to the French, Maltese and Span-
iards. And so with other places and races.

85. Notwithstanding the loss of pioneers, and
even of generations, the process of acclimatization
has ever gone on till the whole globe is
well-nigh occupied by the human spe- Migration of
a Single Race.
cies. To cite one instance, as M. de
Quatrefages describes it, there is the Aryan family of races to which we belong. It is affirmed by
many, though somewhat disputed at present, that
this family, so soon to be divided into many distinct races, started from the mountain district of
the Bolor and the Hindoo Koosh, where the Mam-

ogis still represent the original stock. The region
was one in which the summer lasts only two months,
a climate therefore corresponding to that of Finland.   They descended into Bokhara, and, over-
running Persia and Cabul, reached the basin of the
Indus.   Eleven stations mark the route which the
Aryans followed before reaching the Ganges.   Then
they advanced step by step, sending forward all the
time a vanguard of heroes who slew the Rakchassas,
and prepared the way for future conquests.   That
single family has now accomplished such a migration, that to-day it is in the tropics and in the polar
circle too, reaching from the Gangetic peninsula
and Ceylon to Iceland and Greenland, where the
Norwegians and Danes took the place of the seakings.   It has spread over an immense region of
more or less temperate climate.   And then, when
the era of great discoveries commenced, it distributed its colonies over the whole world, peopling
continents, and displacing more indigenous races.

86. For there were other races before the Aryans,
and there have been others after them, on the same
ground.   Such as had occupied the soil before
them were now mingled together, by the time the
Aryans supervened, and spread over all, reaching
to the western extremities of the European continent, though leaving extensive tracts untouched.
Thus to the North and South earlier races continued to hold their own.   From the time of the coming of the Aryans, documentary history describes

the subsequent progress of nations, and we can read there, in chapter after chapter, of the invasions which followed. In our day, all those nations are found to have intermingled. The characteristics of the original stocks, as exhibited in the peoples now existing, sometimes appear in a crossing of different types, sometimes in a juxtaposition. From the mixture of all, brought together by war and fused by the experiences of peace, the European societies have been reconstructed and formed, as we know them in history, past and present.

87. Similar to this progressive colonization of the Aryans must have been that of the whole human family. Leaving their original centre of creation, the primitive colonists, ancestors of all existing races, marched on by slow stages **Migrations of all the Races.** to the conquest of an uninhabited world. They accustomed themselves to the divers conditions of existence imposed upon them by the North, the South, the East, the West, by cold and heat, mountain and plain. Though many pioneers must have succumbed on the way, yet some survived the hardships of every stage. No matter how hard the conditions of natural environment were, there was always one reassuring feature in the case. It was that the most exposed ranks of the advancing lines had only nature to face after all; and nature, though at times something of a step-mother to the family which she owns upon earth, is ever still a mother. But those that followed, in the vanguard of subse-

quent migrations, had something besides nature to
confront.   They had their fellow-man to meet.
He had walked that way before, and he had settled
down on what he had so dearly bought and now
considered his own.   Every form of selfish nar-
rowness, local, provincial, national, barbarous or
civilized, arrayed itself on the wrinkled front of
grim-visaged war to withstand the invaders.   Thus
it came to pass, that, while the conditions of acclimatization remained the same for the subsequent
migrating sections of the family, the colonizing
movement became much more embarrassed.   The
story of its progress, whether historic or prehistoric, is war.   The only true enemy of man has
been man.

88.   Yet if, under the aspect of political relation-
ship and the mutual adjustment of boundaries mine
and thine, contact between man and
man has issued as a rule in warlike strife,
throwing them back from one another in
the shock of battle; under another aspect, that of
anthropology, the same contact has for the most
part resulted in quite the contrary effect.   Man's
conquering instincts, however brutal or ambitious,
have never prevailed entirely over his social instincts, whether these latter were just as natural and
brutal as the former, or as supernatural and refined
as Christianity has made them.   By the refinement
of Christianity, I mean such a state of social virtue
as the Rev. R. L. Everett, an English clergyman,

Instincts of
Sociability.

has just described, when he says of a neighboring country that "the people live there crowded together in poor cabins, and thus necessarily are in the way of temptation, and the island is full of inflammable material and dangerous situations; yet it is the purest land under the sun;" and, he goes on to say, even in the wild outburst of 1798, it is admitted on all hands that not an outrage in this respect was committed by the rebels. This is the refinement of Christianity in the matter of the social virtues. The brutality of nature is instanced in a whole territory which is excluded from the American Union, because of the opposite state of things; or in a whole population of immigrants, who are likewise excluded from the ports of California. In either extreme of virtue or vice, and in all the grades between, contact among mankind, even when rough at first, always ends in some degree of gentle fusion. *Simile simili gaudet.* The distinguishing characteristics then of pure races are fused and merged in the formation of mixed races. It has only been the barriers of deep rivers, of mountains and of sea, that have kept races by themselves, have made them develop in all their native strength, and add the adornment of their own peculiar variety to the unity of the human species. But not even these barriers, nor those of racial enmity, or political antagonism, have availed to bar out the social instincts of humanity.

89. The multitudinous shades of black and

white, of yellow and red, show how the fusion of race with race has proceeded in various countries.
And, as it goes on farther, the distin-
Blending of the Races. guishing characteristics become less and less distinct. While new tints of special characterization are added on, there is no discerning any longer what particular varieties of modification underlie the latest; or which is the latest, or in what order they have been added. This second nature, which has thus overlaid itself upon the radical, common human nature, is not superficial, like the tints of a wall. It is, as we have already seen, physical, physiological, anatomical (No. 55–63). So that all the experiences of a race, in the past, have gone towards forming it as it is at present. Who could now analyze the Anglo-Saxon, or the Celt; or distinguish nicely the elements of the New England Anglo-Saxon as racially different now from his English cousin whom he left on the other side of the ocean, nearly three hundred years ago? So too, a century hence, the Irish American and the German will have undergone the physical modifications, which the climate and other conditions of the New World are imposing upon them.

90. Here we see how the general fusion of all nations is tending towards a general similarity, in
The Similarity Resulting. which opposing characters are merged. The growing facilities of international communication are tending to develop in all men the same views, inclinations and inter-

ests. What Christianity has long been doing by
the formation of a universal Christendom, with an
elevation of thought and morality, which makes
all alike to be children of one father and brethren
of one another, the parallel process of opening up
the resources of nature for the service of men has
been doing for their physical amelioration, from the
clearing away of the woods round the monastic
centres of the middle ages, as Hallam describes, to
"the putting of a girdle round the earth in forty
minutes," by the electric current of to-day. And
a natural reunion is being effected, after the long-
standing separation and dispersion of the family;
preluded indeed by many a partial reunion, but
never so complete before.—I said, on beginning
this survey, that anthropology supplied a view of
its own, supplementary to the ethical and intellectual history of man. We see now for ourselves that
it supplements also the history of Christianity and
the supernatural development of man.

91. While the old races will not henceforth be
marked so sharply, new ones cannot be formed so
easily. The spread of useful information enables
men to protect themselves more effectually against
the further inroads of climate. The ready facilities of transit from one climate to another aid in
accomplishing the same result of self-protection
and immunity from further racial change.

92. Nor are certain moral agencies wanting to
merge all into a close unity. In the bosom of races

already mixed, an unconscious selection is ever trending upwards, in a favored direction, to-
ward what is conceived as the superior
race. The half-breed will not marry
down into the lower stock from which she
partly sprang, but up into the higher, which alone she
cares to recognize. And it is not only the superior
race which nature favors thus. Among superior
races she discriminates in favor of the more vir-
tuous one. Here a word of Holy Writ seems to
be verified in the mere natural order of fecundity:
" The brood of the wicked shall not thrive." For
the prevalence of vice, either by direct malpractice,
or by the mere deterioration of the parents' constitution, is tending to obliterate races; as it obliter-
ated the Romans, masters of the world, and filled
their places with a healthy infusion, first of freedmen, and then of barbarian hordes. Without recurring in modern times to freedmen or barbarians,
nature is avenging herself upon races which had a
right to be considered prominent in many a native
quality and talent. They are dying out right under
our eyes, themselves lamenting the fact. And she
is calling upon nations, not inferior in other respects, but superior in Christian purity and virtue,
to come and engraft themselves upon the decaying
stock, or to supplant the race, when the measure of
its iniquity is full.

93. All this seems to foreshadow that, as the
Greek world of old fused many little tribes into one

The Dis-
crimination
Practised.

polity; and the Roman empire, many nations into the concentrated majesty of the Roman peace; so the simple unfolding of **The Man of the Future** one Providential design, in the great world at large, is bringing about the final fusion of all peoples in the grandeur of a more exalted unity; and, if leavened with Christian virtue, of a deeper and more sublime peace. It will be more strong than that of the Romans, more intellectual than that of the Greeks. It will exhibit every tribe and people and nation under heaven, coming back to the reunion of the family, with the marks and scars perhaps of many a weary wandering and bitter conflict in the past, but also with the personal shades and forms, as well as the social qualities and beauties, developed by every soil and climate, and unfolded by the side of every flowing water, or in the bath of every limpid air which breathes under God's azure sky. They bring back into the common fund of our original nature a positive contribution of their own; and that is the second nature of their acquirements, of their characteristics and endowments, which they have travelled the whole world to make their own in the past, and which they leave as a heritage to the man of the future. Every thing which they thus made their own in youth, they cannot but retain in their old age. Fifty years in the vigor and susceptibility of our family's early life could suffice to develop, and that to its maximum effect, what fifty generations sub-

sequently can neither moderate nor intensify. The
wine, once poured in deep enough to assume its
full color, will not change that color, whether the
infusion be suspended, or be increased tenfold.
And the vase, once imbued with the odor thereof,
may never again lose it. The diversified races of
mankind, imbued with all the qualities of all their
antecedent experiences, come back to a final reunion, and contribute a blended character, seasoned
and "compounded as with all the arts of the per-
fumer," for the personal patrimony which they be-
queathe to posterity.

94. Nor does this mean that the same posterity,
which the Divine Providence over nature was ever
looking forward to as the accomplished
heir of antiquity, will subside into a
sameness as of all characteristics neu-
tralizing one another in a dun-colored
mediocrity. Active characteristics of the human
mind and heart and body do not neutralize one
another; they go to balance a character, and perfect one another. Besides, as long as there are
poles to this globe, with an equator and two hemi-
spheres so different between; as long as there are
continents and islands, mountains and plains, there
will still be sufficient distinction of characters,
physical, intellectual and moral. But that only
vindicates for mankind the more perfect unity of a
Providential design, as being so conspicuous in a
still abiding variety. Unity and variety! As these

Posterity still varied and beauti-ful.

make up every other kind of beauty, so do they constitute the family comeliness of the species called Mankind, which is quite at home in its own world.

———◆———

95. Let us recapitulate now and conclude these two chapters. The society of mankind, in its natural groups and general formation, is the sub-ject-matter of the science called anthro-pology. Going back from where we stand towards our origin, we find ourselves reaching in different places, at points in the past unequally distant from the present, that state of dim obscurity and uncultured ignorance, which afflicted certain outlying margins of the human species. This state has been called the prehistoric, that is, prior to history. And it is so, with refer-ence to local and partial history; not however as referred to the universal history, which is handed down to us with the documentary evidence of the Mosaic narrative. Prior to the point of history with which that narrative begins, there is nothing prehistoric.

Recapitula-tion. Results of Anthro-pology.

96. Away from the margins inward, at the cradle of the human family, archæology with its clearest renderings of its own records is but a feeble commentary on the distinct and articulate history consigned to Babylonian bricks, Vedantic books, Oriental annals generally, and most of all, because the clearest of all, the narrative written by Moses.

These documentary records are endorsed, at every principal chapter and page, by universal traditions, by the science of linguistics, and by all the resources of science which the subject admits of : they concur to the effect that Noe was, and Assur, and Nimrod, and Misraim, and Adam, and then— what ? Not the subject of a prehistoric anthropology, that is certain. For there was no man at all then; and so there is no science about him, either historic or prehistoric. History brings us back to a first man; and, before him, anthropology, or the science of man's life upon earth, has not begun to be.

97. But when it has begun to be, when as a science of to-day it finds where its subject, man, began to exist, it traces with diligence the history of our species, as we left the cradle-land and wandered through many vicissitupes, even to the division of races. It contemplates us with color changed, face and form and brain altered, everything gone from its original style and feature, except that specific sum of characters by which a man is a man all the world over. In many lands our wanderers have stood. Now, when every shore is explored, no single one is found without the signs of our labors. The long tale is told of liberty used and of liberty abused. Traditions coming down to us by many an avenue, like an old melody never lost, sing of a better time that was, and of a supernatural state which was and which ceased to

be—of a sin, and then a fall, and then, after many days, of a future better state that is to come. Thus, forward and backward alike, natural science bears us to the supernatural. In the retrospect and in the prospect, the human ever leads to the divine, as the waters to the sea.

BIOLOGY.

CHAPTER III.

SPECIES; OR, DARWINISM.

98. We quoted the German Professor Virchow as saying at the Congress held in September, 1887, at Wiesbaden, that the transformation of one species into another was a theory so destitute of proof, as to find no place as yet in the domain of science. The most ardent partisans of transformation, like Haeckel and Vogt, are forced to admit separate origins for the principal classes of species. With this utterance of the Berlin professor others agree, while still professing sympathy for the theory of Mr. Darwin on the descent of species. Thus Professor Huxley, in his recent notice on Darwin in the Proceedings of the Royal Society, practically concedes that natural selection by the survival of the fittest cannot be the only or chief law to explain the origin of species. On other occasions, as in his work on Man's Place in Nature, he affirms that the necessary physiological proof, which regards the sterility and fertility of species, is still wanting to establish Darwinism.

87

99. However, no chorus of dissentient voices seems to affect the almost unanimous popularity which this theory enjoys. Something more than argument is needed to disabuse men's minds; if indeed they were ever mistaken on the subject. Men who know enough are not simply misled when they adopt an ill-founded theory; and, as to those who do not know enough to judge for themselves in the matter, neither are they always deceived when they follow the leader. There is something besides logic that may be interested in a question. The distinguished poet and moralist, Aubrey de Vere, remarks upon the subjective difficulties of men with regard to religion, that the logical faculty is but a part of man's understanding; and his understanding again is but a part of his whole being. Poetical criticism affirms the same in its own sphere; and rhetorical analysis does likewise. So there is quite another part of his being than his logic that can be appealed to and challenged by what makes a theory popular or otherwise. A man may know nothing at all about science, " not even the first word," said an eminent anthropologist, the Marquis de Nadaillac, at a scientific congress held last spring in Paris; but, without learning even the first word of science proper, he can catch well enough that such and such a theory means " the denial of creation, the denial of a Creator. God is the ancient regime;" and a theory which promises to change this old order of things appeals to man

on a more sensitive side than his logic. But then the logic of the mind may somehow be brought over to the cherished opinion of the will; as so many a fond wish is father to the thought. Thus the comic poet said of old:—

I brought my mind to be of my opinion !

100. We for our part, as philosophical critics, are not free to treat this question otherwise than from the side of logical reason. Still, it helps the setting and poising of a logical mind to understand the moral temper which it has to meet. The temper in question, which so readily accepts of a theory unproved and disproved, is that disposition called naturalistic or materialistic, which has spread over the whole body of society, biassing the judgments of some, enlisting the affections of others, dominating the profane sciences, and not scrupling to interpret sacred science. It invades legal ideas, and, from the clinic of the physician, it passes over to sway the verdict on a criminal. It touches the morality of business, as well as interests political and international; and it is shaping every department of letters.

101. Under the scientific form of Darwinism, around which, as a quickening nucleus, the whole theory of evolution has gathered, we find this naturalism or materialism acquiring such an ascendancy that every science now pays tribute to it, and to the theories which, wrongly or rightly, are taken

to be an expression of it.   Darwinism, or evolution
in general, determines the cast and phraseology of
those sciences which we are now reviewing, such as
botany, zoölogy, anthropology, geology.   In chemistry, it introduces the idea of meta-elements to
construct an ideal passage for the evolution of one
simple element into another.   In astronomy, the
evolution of the stars runs its course apace.   In
philosophy, after reducing the whole of it to the
terms of evolution, Mr. Herbert Spencer lays the
bases of evolutionary morals.   His disciple, M.
Letourneau, then evolves marriage and the family.
M. Duval finds a comparison to institute between
the evolution of organisms and of languages.   History is interpreted by evolution; and Scripture even
is quoted by Mr. Bancroft to the same purpose,
that ever a new messenger of " the Infinite Spirit
moves over the waters; and the ship of destiny,
freighted with the fortunes of mankind, yields to
the gentle breath, even while the beholders still
doubt whence it comes, and whither it will go."
All this falls aptly into the mould of German trans-
cendentalism, *das Immerwerden*, which M. Renan
translates becomingly as *l'eternel devenir*.

102. Some ardent admirers of Mr. Darwin have
claimed that he has given a scientific and demonstrated reality to ideas which were conceived, indeed, but not established, by his predecessors.
His system is a dogma of science now.   It is the
only scientific theory, if we can rightly define

science as " the elimination of the supernatural (that
is, God) from all explanation of natural things,"—
leaving to the First Cause liberty to walk about the
poles of heaven, but keeping the earth to ourselves.
There have been predecessors to Darwin both in
the line of natural science, and in that of political economy.   The English political school of
Hobbes, Adam Smith, and Malthus treated mankind as originally savage, then struggling with
one another for existence, then striving upwards by
the survival of the fittest and various compacts to
the present state of civilization.   The evolution of
civilized man, which they assumed for the purpose
of political economy, is now a favorite common-
place in natural science, and was the subject of
Major Powell's address last spring when retiring
from the presidency of the Anthropological Society.
But it is the evolution of brute animals into man
which forms the final purpose and object of Darwinism, by showing that one species can change
into another.   This we call Specific Evolution.
It is our subject at present; and it derives its
interest from the possibility, that, if one species
ever evolved out of another, the ape may have
changed and evolved into man.   It is the same consideration which extends a similar interest to the
question whether living things, or living cells, ever
evolved from non-living elements; a development
of life from non-life by spontaneous generation,

which comes under the head of Organic Evolution. We shall treat of that when we come to Cells.

103. Accordingly, addressing ourselves to the present subject, let us see for ourselves whether Darwinism is the scientific theory which it is claimed to be by some; or whether, as others affirm, it has only been more gratuitous and extravagant than any other powerful superstition has ever shown itself to be, when in the lightest of moods.

———◆———

104. While treating of human races, we had occasion to define the term, Species (No. 50–54). It
Species, Race. was necessary for the purpose then; but we were only anticipating the topic now before us. This idea of species is the prime one in the present question; yet it is the one for which Mr. Darwin cares least. The other elements on which he has enlarged, at a length and with an emphasis quite out of proportion with their value, are easily handled and located, if this idea of species is caught. But there would seem to be a great interest at stake, in keeping the idea obscure: "See what a surprising number of forms," he says, "have been ranked by one botanist as good species, and by another as mere varieties." So he speaks at the beginning of his book on the Origin of Species; and instead. of doing what all logic, science, and common sense demanded, define and clear up by induction and deduction what a species is and must be, he takes that confusion which he finds hanging

about the term as his point of departure; and slip-
ping over the momentous bearings of the one definition, " species," on which the whole value of his
book, the Origin of Species, must depend, he gives
us a momentous book in consequence. The necessity of this confusion for the maintenance of his
system is so well understood among some followers
of his, that, as M. de Quatrefages classifies them in
chapter VIII of his Unity of the Human Species,
some among them cry out, it is useless to go about
discovering what species and race are; others com-
plain that naturalists have so many definitions of
species; others again that people who define species
and race, in the accepted way, are running about in
a vicious circle of logic; others, that in general
there is a want of precision here. All these classes
alike feel that, for the purposes of Darwinism, a
clear definition of what we are speaking about,
when we talk of species, is cutting the ground from
under their feet, is an impertinence in their line of
science, and leaves the graceful curves of thought
and observation, which Mr. Darwin knows how to
describe, without any logical origin to start from—
a performance usually considered suicidal, and
therefore justly eyed with disapproval by them.
So a clear definition finds no place with them.

105. Indeed the point from which the Darwinian
series of speculations first took their rise, and since
then are

Wont to roam from shade onward to shade,

may be cited as an illustration of the essential
obscurity, in the shadow of which the system lives.
He observed some 150 races of domesti-
The Origin of
Darwinism. cated pigeons, developed into such differences of form, under the action of
artificial breeding, or selection, that had they been
found so in a wild state we should have been compelled, he says, to make three or four genera of them,
and still more species. Now, at the same time, he
observes that the wild rock pigeons, the original
stock of all those domestic races, differ only in shades
of color. These observations of Mr. Darwin would
show a logical mind two things; first, how much more
effective is artificial breeding, or selection by men,
than the mere operation of natural conditions,
which he calls natural selection; and secondly,
that the test of species could not be in color, or
form, since these 150 races differed ever so much
in color and form, and yet were one and the same
species. So would a logical mind infer. How did
Mr. Darwin proceed from these identical observations of his own? In quite the opposite direction,
and in the teeth of his own results. First, he declined to confine himself to any definition of species,
because there is so much confusion among botanists
and zoölogists about the use of the term. Secondly,
he chose to surmise that, since artificial selection
had done so much, as to discriminate 150 races in
a single species, when natural selection had done so
little, as to keep its original rock pigeons not only

in the same species, but pretty much in the same race, then perhaps all the great species of nature, divided off as they are from one another by vastly more than differences of race, might have come about by natural selection!—the great fact staring him in the face, that even artificial breeding had not discriminated any one of those 150 races into a species different from the rest! In that logic Darwinism originated, and its subsequent career has never thrown off the conditions of its birth.

106. For those who are interested in the play of a logical faculty, it may be useful to take note here of several characteristic traits, the illustration of fully half-a-dozen sophisms or fallacies, as they are called in logic. He starts an hypothesis to explain what he has never seen occur, the formation of species out of mere varieties, or races. So the point to be explained is gratuitous; and the hypothesis assumed to explain it is equally so; nay, more so, for the natural selection, which he invokes to explain it, only tells, according to his own observation, in the opposite direction. Still he sets up his hypothesis that natural selection had formed different species when artificial breeding had formed only different races. Here he commits a fallacy in analogy, likening two things to one another there where precisely they are seen to be different. While one process of breeding is seen to select and form races,

Sophisms:—
Analogy,
Equivocation, Comparison,
Erudition,
Hypothesis,
Begging the
Question.

and the other forms none at all, he argues about
both alike as being each a process of selecting, and
calls both " selection." Unless it be that he commits the error of mere verbal equivocation, when
he uses the same term, " selection," for things so
entirely different,—one process of breeding being
but a chance result of chance combinations, the
other a result aimed at by man through combinations designed to reach it. In either of these
alternatives, you have a sophism illustrated. Thirdly,
he commits the fallacy of an inverted comparison.
For, comparing the two kinds of breeding together,
he should have concluded that, as man's artificial
selection had not formed different species, still less
could the blind, mechanical operation of nature do
so. This is the argument of likelihood, *a majori ad
minus:* what was the more likely did not occur,
therefore neither could the less likely. In the face
of the obvious facts, he infers the opposite, that
perhaps blind, natural selection brought all existing species into being. Fourthly, having entered
on this path, he will proceed on it with all the
pomp of abounding observation and experiment
pleasantly described, which in the premises is but
another fallacy, that of misguided and misleading
erudition,.

That leads to bewilder and dazzles to blind.

Whatever he says henceforth will have the air of
induction, to establish the point to be proved. He
will describe facts of natural history; he will por-

tray the action of the environment, and the conditions of life; he and the Darwinians will show that a number of facts established in the comparative study of organic species are just as they should be, if species were descended from one another. Hence, fifthly, one hypothesis will be made on demand to fit into another; and a huge system of hypotheses will grow round—what? to prove what? He has not defined in nature the ground of the original question—what a species is ? and whether there is a single derivative species existing, as a matter of fact, to lend some color to the question and hypothesis which he suggests! So that it is a system of gratuitous hypotheses, gratuitous because uncalled for by any facts, and gratuitous because unproved by the subsequent hypotheses. Wherefore, sixthly, it is not surprising that, besides being gratuitous, the system should elaborately be only begging the question which it pretends to prove, and begging for new hypotheses to prove it. The fecundity of this school, in devising something newer still to prove what is very new, seems a happy illustration in logic of what I quoted before, *das Immerwerden des Neuen*, "the ever becoming of the new;" and indeed it is a process which, if it once has a reason for beginning, need never be checked by any sufficient reason for ending. Every day we are witnessing new phases of its evolution, *l'eternel devenir*, in logic as well as elsewhere. And as to the materials on which Mr. Darwin's erudition expands at large,

variability, laws of correlation, permanent characterization of species, use and non-use, struggle
for existence, survival of the fittest,—some small
portion of them are substantial matter of observation, others are plausible, others are not so.   But
they prove nothing in his theory, since logically
there is nothing to prove.   Hence the pertinent
remark of the German scientist, Wigand, that Darwin has done nothing but " wrap a theory up in
facts;" leaving us, I suppose, to unpack the bundles,
eliminate the smuggled theory, and reassort the
facts as they should be.

107. He was quite alive however to the difficulty
arising from the true definition of species (No. 51,
52), that by which the physiological test of it is determined, and which we shall take up at once.   How
he endeavors to elude it, we shall then record.

108. To resume our definition then, the term
species, like other terms such as genera, tribes,
families, orders, might be taken to designate classes,
assorted for the sake of convenience.   We defined
it on a former occasion most strictly, following
therein the advice of Paul Broca, as quoted approv-
ingly by Otis T. Mason of the Smithsonian Institute: " Let everything have a name; and let it have
only one; and let that name designate only one
thing."   We noticed before how the fallacy of
equivocation was committed with this very term,
species (No. 51).   Following M. de Quatrefages, we
formulated a definition or description to this effect,

that a species was a collection of organic individuals more or less resembling one another, in their external aspect, or internal structure; productive in their unions among themselves, so that they perpetuate the same collection in nature, by generating other individuals of the same kind; and one of the consequences thereof is, that originally all can have descended from one primitive pair, identical in kind with themselves.

109. In this definition you have divers elements. There is that of likeness, whereby they resemble one another more or less. There is that of filiation, whereby the members of the group, or the posterity spoken of as continuing the same collection in nature, are the offspring of their predecessors in the same class. There is heredity, whereby the group perpetuates certain qualities, having received them by the process of generation, that is to say, by proceeding as living beings from living beings in the same likeness of nature.

Likeness, Filiation, Heredity.

110. Such classes as these are what the term, species, strictly taken, is meant to designate. Now does scientific induction show that such classes really exist in nature? It does. Nearly one hundred and fifty thousand are enumerated in the animal world alone. The individuals composing any such class may differ in form, shape, size, features; but they remain identical in a certain natural capacity for uniting among themselves, and perpetuat-

ing their kind by fertile generation. This one physiological attribute common to the members of a species reveals an essential likeness among them all, one deeper than proportion or structure, than morphology or anatomy. It may not be patent to the eye: it may not be discernible with the help of the measure or scales. Yet consider, first, the parents with respect to the offspring. Even supposing the likeness between them is not very apparent, still it must be there, according to the law that "Like begetteth like," and "No one gives what he has not got." If the parents give, it is what they have got that they give; and in every order in which they give, in anatomy, morphology, physiology. The resemblance passes down from the beings producing to the beings produced, as the old philosophical definition of generation clearly enunciates: "The process of a living being from a living being, unto a likeness of nature." Consider, secondly, the parents themselves. Suppose that previously the likeness between them was obscure. Yet, from the moment they produce an offspring common to both, that offspring is like to them, and they must be like one another; according to the mathematical principle, that things which are equal to the same are equal to one another; and this, albeit nobody knew of it before, perhaps because their color or stature was different, or their origin, their antecedents or concomitants generally. They were never different species, if they are found to be capable of fertile unions. Science may have mistaken the case, as science has yet many a mistaken case to rectify. Not so nature.

111. By the very necessities of the case then, all the individuals of one lineage are endowed with an essential identity, as a family heritage. It shows itself first and radically in the physiological function we speak of, that of mutual fertility for perpetuating their kind. Secondly, it does not fail to reveal itself in form or structure, notwithstanding many variations. For, though individuals and races are quite susceptible of these variations, whether minute, as Mr. Darwin generally supposes, or abrupt, wide and even monstrous, it is noteworthy that such as diverge most from the middle, normal type of a species, are the least stable in maintaining themselves distinct: they tend to fall back and resume the more ordinary type. Hence, as we saw in the observations made by Mr. Darwin upon pigeons, the races left to themselves in a wild state did not vary much; they differed only in shades of color. That was the operation of natural selection, or of the native conditions of life; while it was only man's interference and watchful supervision which availed to make races diverge widely, to keep them apart and unmixed; so that, as Mr. Darwin says, the 150 races, differentiated by man's artificial selection, would in form and appearance claim to be classified, not only in different species, but in three or four different genera. Some species show less

plasticity than others in yielding to selection and forming varieties, as Mr. Darwin observed in the case of the goose. But, in general, so great is the power of scientific breeding, or artificial selection, that, as Lord Somerville remarks of the sheep-breeders in particular, "it would seem as though they had chalked out upon the wall a form perfect in itself, and then had given it existence."

112. What we are saying here is so fully borne out and developed further by Dr. G. Romanes, that we cannot do better than refer here to his falling out with Mr. Darwin. This extreme Darwinian had, so far as we ever observed, only turned the whole force of his scientific journal, *Nature*, with all the tactics of which he was capable, towards maintaining, defending, propounding, explaining Darwinism, with a zeal more than discreet and enthusiastic, rather fanatical than scientific. Yet about a couple of years ago he proposed a new theory of his own, called Physiological Selection, saying querulously of Mr. Darwin's theory: " Natural Selection has been made to pose as a theory of the origin of species, whereas in point of fact it is nothing of the kind." He says there are three cardinal difficulties, which stand in the way of natural selection being considered a theory of the origin of species. Reduced to a brief compass, the difficulties stand thus:—

Three Objections to Natural Selection.

113. The first is with respect to mutual fertility; a huge difference exists here between species and

racial varieties. If species are only the modifications of organic types produced by natural selection, how have they come to be
1. Sterility
of Species.
mutually sterile, when even greater modifications of such types, produced under our very
eyes by artificial selection, do so generally continue
fertile? There seems to be only one answer, that
the said natural species were never produced by
natural selection.

114. In the second place, you suppose that uniform conditions of existence may have acted for
long periods of time on the physiological
2. Time and
other
Accidents.
system of certain varieties, so as to make
them mutually sterile, and thus create
our present species. Such a supposition will not
stand. This sterility, to be of any use in the theory,
would have to arise at once when the variety or
race was just beginning to develop; it would have
had to protect the variety from intercrossing with
the original form, which otherwise would swamp it
forthwith. So we are to make the gratuitous supposition, that sterility arose all at once, at the precise moment it was wanted, and just by chance.
If you will insist upon supposing again (hypothesis
upon hypothesis, which is true Darwinism!) that
uniform conditions of life happened to act upon a
sufficient number of individuals, during the same
interminable periods of time and ever in the same
way; and effectually guarded the new development, and happened to prevent intercourse with the

original form, and so forth, bringing about the actual sterility in some way or other—all this is such an assumption for you to make, that, if you must make it to meet the difficulties, you only increase the difficulties by doing so; and the whole theory becomes as desperate as the assumption, for which " even the chapter of accidents has no room."

115. Thirdly, utility, or the consideration of fitness, is the great mainstay of the natural

3. Utilities.

selection theory, and of the survival of the fittest.   But pray tell us, what utilities are to be found in the differences between many of the species?   Is there any utility about them?   They show on the surface of things only small and trivial differences of form and color, and meaningless details of structure.   If natural selection proceeds by the survival of the fittest, there is very little fitness, and still less of the fittest, apparent in such distinctions.   You suggest that, after all, these distinctions may be of a disguised utility.   But that is to reason round and round in a circle—the *circulus vitiosus* of logic.   These species survive because they are the fittest.   And why are they the fittest?   We do not know, except that otherwise they would not survive! " It is certainly too large a demand on our faith in natural selection to appeal to the argument from ignorance, when the facts require the appeal to be made over so large a proportion of instances."   So far Dr. Romanes.

116. To these words we have only to add, that

if " rudiments," as Mr. Darwin calls them, are to be claimed as a proof of natural selection and survival of the fittest, we shall yet see how very little any principle of utility could have operated to produce such utterly useless parts as rudiments are seen to be (No. 129-131).

117. And Mr. Darwin,—how does he forearm himself against the inevitable and insoluble difficulties which loomed up before his theory even in those times, and which now are made to stand out clearly around the whole horizon of science, thanks to the reaction against his aggressive speculation? His defence is quite characteristic. I call your closest attention to it, not because it is hard to catch, or rare, but because it is precious. He speaks thus in his " Variation of Plants and Animals:"—" Since species do not owe their mutual sterility to the accumulative action of natural selection (so he granted the point beforehand), and a great number of considerations show us that they do not owe it to a creative act (this is an argument *ad odium*, to excite the anti-religious prejudice against another theory), we ought to admit that it has been produced incidentally during their gradual formation (this is begging the question which he ought to· prove, *petitio principii*), and is connected with some unknown modification of their organization." (This last is such an ineptitude, that, instead of dignifying it with the name of any

Mr. Darwin's Manner of Answering.

sophism in logic, we may prefer the term of Mr. St. George Mivart's, and call it " a puerility.")

118. Elsewhere, not knowing how to deal with the fact that mongrels or cross-breeds between races are always fertile, while species remain so sterile, a fact of observation as big as the world, an argument of complete induction which we have ascertained, and we know, Mr. Darwin placidly remarks: " We do not know whether the mongrels of wild races may not be sterile."

119. *Posito absurdo, sequitur quodlibet,* said the old axiom: " Admit an absurdity once, and anything will follow." Make one absurd supposition, and there is no freak of logic which is too much for you, no figure or curve of fancy which you cannot execute. Put up one hypothesis which will not stand, and, as in other questionable avocations, your genius and the best of memories will be taxed for twenty others more and more deftly contrived to keep the first one up, some way or other.

120. " Unknown!" and therefore we must grant it! Or, as he says of the missing links in the chain of beings which geology should yield up, but does not, " They may now be in a metamorphosed condition, or buried in the ocean." It is like Professor Haeckel's assurance when, in the gravest of arguments, as that of biogenesis, he requires such and such admissions to be made " for the most weighty general reasons;" or that the existence of some unknown animal, the sozoura, must be granted, since

the proof of its existence arises from the necessity of its being there! The philosopher, Mr. Herbert Spencer, besprinkles his pages with this kind of formulas. It is all begging the question, assuming what has to be proved. But then the assumptions once made are taken as the basis of subsequent " demonstrations." And here, with Mr. Darwin, his point is not proved; it is positively disproved. . The answer is: No, let us say we do not know anything about it, and therefore grant it!

121. Hence we have a museum of paralogisms and sophisms in the modern school of science. We may note four or five more, in addition to the half-dozen already instanced. In the first place, chance, possibility, or their own personal convictions are taken

Sophisms :—
Chance,
Induction,
Post hoc,
Non-causa.

by these scientists to be convincing reasons for others. Chance is an essential element in Darwinism, for this theory excludes anything like a principle of evolution upwards; it posits only mechanical adaptation to whatever conditions of environment may occur. Possibility, a term oftentimes denoting the impossible, is taken as a demonstration; and, once used, it is thenceforth kept before the mind by a suitable term; as we now see the term " evolution" on every side, or the term " simian," ape-like, or of apish affinities; as if these things had been proved once for all to be facts, and now needed only a term to recall them. This is only taking for granted the thing to be proved. Or, secondly, the assumption

of having proved the point is borrowed from another
side; when, finding things to suit their anticipations,
the scientists argue that evolution must be true
because the facts are shown to be in entire accordance with it.   This is what Professor Huxley calls
the demonstrative evidence of evolution.   And it
would be so if all the facts were shown to be in
perfect accordance with the theory, as all the celes-
tial phenomena are in keeping with the Copernican
system, which is consequently no longer an hypothesis but a thesis.   But not so, if only some facts
agree, while others and many others contradict the
theory.   Then the hypothesis is disproved; and to
take it as demonstrated on the strength of some
coincidences is a fallacy of sophistical induction,
like that which we noted in geology (No. 16).   And,
thirdly, here comes in the use of another sophism,
when those very facts, cited as being in perfect accord with the theory, are so indeed, but are in
entire accord with a different and contradictory
theory.   Thus, Professor Huxley finds his demonstrative evidence of evolution in the series of fossil
horses, representing different stages of evolution up
to the recent, modern form.   In the fourth of his
American lectures on the subject, he shows the
fossil forms, orohippus, mesohippus, miohippus,
hipparion, pliohippus, all verging in the same direction, and finally the series terminates in the modern
horse.   This is just, he says, what evolution would
require.   One may answer: But it is just what

Leibnitz and Linnæus would require, who held the principle that *natura non facit saltum*, "Nature is continuous." It is also what St. Thomas Aquinas and Aristotle require; for the scholastic principle is, *supremum infimi attingit infimum supremi*, " The extremes of different orders touch one another." These philosophers did not sustain evolution in the sense of Haeckel or Huxley; yet they can claim for the genuine philosophy of evolution the argument drawn from the series of horses, and return thanks for it to Professors Huxley and Marsh. When therefore, at this late date, a scientist claims the argument for his new and latest theory, he is reasoning *post hoc, ergo propter hoc*, on account of evolution because the discovery of the fossil forms happens to date after the theory, and to fall in with it. That was a *post hoc, ergo propter hoc* argument, which Lord Bacon pleasantly records as used by the house-fly; perched on a chariot wheel, which was whirling along and raising clouds of dust, the fly said complacently: "See what a dust I raise!"—a sophism quite familiar to us nowadays. Fourthly, not very different from this is the line which Mr. Darwin follows when he fills his books with descriptions and narratives of what nature does. And, having described the facts, he assumes that he has explained the cause; and in the light of his theory he calls the description, "natural selection." There is no harm in calling your description anything you like; just as Mr. Spencer is free to define evolution any way he likes. But 'it is not well to give the name of your definition or the name of your description to nature herself and her processes, if the facts do not answer the name. That would be the fallacy of assigning the wrong cause or no cause, instead of the right one, *non-causa pro causa.* But enough of logical fallacies for the present.

122. If any one should like to see the working of a true cause and its true effects, let him take that idea of species (No.108) which we have reported as truly drawn from the facts by a process of strict induction; and, reversing the process to one of deduction, let him look out upon the order of nature, and see the organic world resulting as in exact accordance with that idea of species. He will observe three philosophical corollaries proceeding therefrom, and going far towards accounting for all the unities and varieties discernible in organic nature, as well as for many of the errors discernible in natural science. In the first place, he will see that descent by generation implies fertility in the parents, as well as an essential similarity between them. It also implies similarity between the parents and the offspring, *simile generat simile,* "Like begetteth like." So the offspring will be fertile among themselves; since that is included in the essential likeness to their fertile parents. Thus then descent implies continued fertility; and fertility secures continued

Deduction :—
1. Uninter-
rupted
Descent.

descent; and the species goes on indefinitely throughout time, extending over the globe. Retrospectively also the species must have subsisted thus with respect to likeness, to fertility, and uninterrupted descent, from the time one primitive pair produced this collection which originated in them. Whence did that pair originate? Darwinism does not say, beyond suggesting that there were four or five original stocks from which all species started. Evolution, in the wider sense, must go far beyond. And though, as we heard before (No. 98), even such ardent evolutionists as Haeckel and Vogt are now reduced to admitting separate origins for the principal species, still genuine evolution must go on boldly to affirm that all organic life sprang from non-life; and, if necessary for the purpose, it must affirm spontaneous generation. But of that after a while (No. 154–160).

123. In the meantime we must take note of a second corollary, with respect to the essential likeness, which, as the axiom says, is the ruling attribute of a species, *simile simili gaudet*, " Like rejoiceth in like." That 2. Unities and Varieties in Nature. likeness consists first and foremost in a unity of function, by means of which it is inclusively fertile and exclusively sterile. It sets up specific barriers, within which it includes all individuals and varieties of the one descent, while it excludes all of any other descent. If we look out over the organic world at large, we see these specific barriers maintaining all

the unities and varieties which characterize organic
beings, and making up of them all, in their species
and genera, a beautiful world; inasmuch as, exclusively viewed, the species are different, are outside of one another; precisely because, inclusively
viewed, their physiological characters remain constant, and keep each just what it is. There is in
them what Mr. Darwin chooses to call a "law of
permanent characterization," keeping a species permanent in its characters.

124. Indeed, beyond the species, throughout all
the genera which comprise them, there is visible a
more general likeness, a still wider unity, in function and form, in the elements whether anatomical, chemical or mechanical,—a unity so marked
and express, that the schools of evolution, struck
by the analogies throughout all nature, are prone
to see nothing else there but a solid unity, without
the varieties; or, if they will see the varieties, pay
such exclusive court to the unity as to reason that
all must have come from one, if they are so bound
up in one plan. So far the reasoning is correct;
because, as the axiom says, *multa non reducuntur
ad unum nisi per unum*, "Many things are not
brought to unity but by a unit,"—a unit in the
design and in the designer. But they proceed
otherwise, and infer that all must have come from
one stock, descending thence as in a single family
or species. No doubt, if they were of a single
family descent, they would be bound up in a unity

of likeness. But it does not follow, *vice versa*, that if all are of one make, therefore they are all of one descent. Or else, metal, a stone, a house ought to be of one descent with ourselves, since we all have weight, are impenetrable, and the like. There would be a fallacy in drawing such a consequence.

125. No, the true statement of this general unity may well be made in the words of the Duke of Argyll, on the Reign of Law: " Never in all the changes of time has there been any alteration throughout the whole scale of organic life, in the fundamental principles of chemical and mechanical adjustment, on which the great animal functions of respiration, circulation and reproduction have been provided for." And as to the explanation of such a unity, if it can be given without recourse to a design, a designer, and provider, let such explanation be brought forward. But I think we have seen such an attempted explanation put forward at its best in the course of these twenty-five years, elaborated with the combined efforts which all the schools of natural science have made in civilized countries, and with the help of all the marvellous appliances for observation and experiment now under their control. And what the outcome of their work happens to be we are just examining, with a slight degree of pardonable curiosity.

126. A third corollary to be noted here is this, that you will look in vain for such a definition of

species anywhere else save in a genuine, well-ground-
3. A plea for ed science of biology.   It is not the out-
Genuine come nor the subject-matter of medical
Biology. observation, for this regards primarily the
material human subject, as such, not any species in
its subject.   It is not familiar to the palæontologist, who has nothing else to deal with but the bare
elements or proportions of fossils or old bones.   It
is true that, speaking from amidst the narrow resources of his own specialty, Professor Cope said
to the American Association at Minneapolis: " Biological science is a case of analysis and forms.
What the scales are to the chemist and physicist,
the rule and measure are to the biologist.   It is a
question of dimensions."   But he was confounding
biology with his own department of palæontology,
and so missed the point of the question.   For the
question of species is evidently a matter of life, not
of death.   And, if he meant to apply the rule and
measure to the things of life, we should like to
know the linear dimensions of a live instinct, or
the cubic root of heredity and filiation.   Finally,
species thus described does not come in the way of
entomologists, conchologists, etc., who classify what
they call species by purely external characters, and
treat their insects or shell-fish, even when alive, as
they would treat fossils, which are more than de-
funct.   The question is centred upon a physiological quality, that called reproductive fertility.   And,
compared with physiology, all structure and form

are but superficial, resultant attributes, far from being deep enough to determine the order, kind and beauty of the organic kingdoms. Even outside, in the inorganic world, you will find something deeper than form and structure. You can see it in the crystal, and in the commonest chemical elements. How much more in the unity, totality and economy of a living being!

127. Take, for instance, yellow phosphorus and red phosphorus. To the chemist they are an identical element, though he recognizes that the same element must be in different states. Quite so; for the element as yellow phosphorus is an active poison, while as red it is inert. There must be something here besides what chemistry weighs in its scales; and chemistry no less than biology becomes philosophical, in its endeavor to explain it. Again, why should piperine, asks Professor Tidy, be the poison of all poisons to keep you awake, and morphine the poison of all poisons to put you to sleep, although to the chemist these two bodies are of identical composition?

128. Oh, to the truly philosophical mind what a revelation runs through all nature of a design and a designer playing at all times, playing The Miracu-
in the world, delighting to reveal himself lous in
and give occupation to the children of Science.
men! All that we know in every range and sweep of nature is but the smallest part of what remains to break upon us in number, weight and measure,

in species, form and order. Ah! but miracles! exclaims materialistic science. Miracles! re-echoes naturalism. " True science eliminates the supernatural from nature; it cannot admit creative acts or miracles!" This is all an appeal to prejudice, and is called in logic the sophism *ad odium*. The mention of God creates aversion in the bosoms of some men; and, if a revolution in science means the dethronement of His ancient dynasty, then welcome the revolution! But if in the same persons' minds there is something worse than a sin, if there is such a thing as a blunder, I would beg to submit that here there is a blunder in the facts and in the logic. In no sound theory of the world's material development or evolution, of which some three or four might be sketched, is there any question of creation, creative acts or miracles. After the first creation of matter, there is only the normal action of a Supreme Cause's administration, or government; and that is not creative, nor miraculous. If anything were miraculous in the development of nature, it would be species evolving by descent out of ancestors or elements that never contained them, if the nature of the case admits of no such production. This kind of evolution would be the miraculous in very deed. And M. Ferrière, an evolutionist of the most materialistic type, turns round sharply upon Haeckel for maintaining the miraculous, and the absurdly miraculous. Professor Haeckel had said in a discourse delivered at Paris: " Whoever does not

believe in spontaneous generation (life out of non-
life), admits miracles." Ferrière of Haeckel's own
school answers in his book on Darwinism: "As if
the formation of living beings, by a crystallization
of carbon or lime, were not a *miracle just as absurd
as*" creation—but, after italicizing the absurdity of
Haeckel, he expresses the idea of creation in terms
not unworthy of the school and its ancestry. Pro-
fanity, it would seem, is included by some of these
men among their scientific credentials, to commend
them to their kind. But, if that is only a sin which
they rather affect, there is something which is worse
in their eyes, and they seem unable to avoid it—
that is blundering. Without however pursuing any
such psychological rudiment back to its origin in
their moral structure, I prefer to hurry on to a characteristic argument of theirs in the present question
of species, and with it to finish the definition of
species which has occupied us thus far. I refer to
what Mr. Darwin has called rudiments.

129. It is found that in many an organism there
appear certain local structures quite useless as they
now stand. Man has certain muscles
for moving his ears; but he never moves
them now; they seem to have lost some pristine amplitude, when the muscles might have been useful.
The great finny monster called the whale has imperfect legs, or similar structures of motor significance;
so has the boa; but these beasts never walk now.
And though the young whale in the fœtus state has

teeth, yet the teeth are reabsorbed, and·the whale
does without them.   So too with rudimentary wings
in the ostrich, etc.   Now, says Mr. Darwin, in order
to understand the presence of such organs, we have
only to suppose that some remote ancestor possessed
in a perfect state the parts which are at present reduced to this condition.   Such a supposition coin-
cides with the descent of species.   Hence the existence of rudiments corroborates that theory.

130. This argument merits a remark or two, as to
its matter and form.   Could anything show more
The Matter of clearly that true Darwinism is not a sys-
this Argu-   tem of progressive evolution in any sense
ment.   whatever?   It is only a plan of adaptations, of self-adjustment to circumstances, in any
direction, whether towards better or worse, or
neither way; a blind, mechanical variability, having within itself no principle of development towards a higher life and higher species.   In fact,
such is really the Darwinism of Darwin.   Taking
this instance of rudiments on Darwin's own presentation of them, we find that a locomotive apparatus,
which no one will fail to recognize as very useful
to a whale if stranded on a sand-bank, is lost to
the monster, one knows not why—a case apparently
of mere degeneracy; and yet that unexplained loss
fits in perfectly well with Darwinism.   Certainly
there is no progressive evolution here.   Nor is the
locomotive apparatus lost totally; so it appears
doubly useless, as well in what is gone, as in what

remains, encumbering the organism as it does with the silent reproach of the better times that were! That is not evolution. Why, too, on the Darwinian presentation and hypothesis, should the animal lose its incisors, when once it had them in its mouth? They might be vastly more useful than mere whalebone for much of the food which was yet to come into its jaws. As to the matter then of this argument, there is no upward evolution conspicuous in it.

131. And, as to the manner or form of the argument, there is rather a suspicion of downward evolution in it. He illustrates these obscure, functionless organs, not by what is clear *Its Manner.* and substantiated as matter of fact, but by what is obscurer still, the full legs which no science has ever yet verified in the whale or the boa, the long asinine or apish ears, which no one has yet seen in man. "Ah! but—" a well-intentioned scientist answers—"wherever rudimentary organs exist in one type, they are sure to be found in their normal state in a neighboring type!" The reply to this argument is very simple: it is merely to ask, what does that prove? We do not want these crude premises with only implied consequences, after the style of Mr. Darwin, who wraps up a theory in crude facts, and leaves it there. We should like to know distinctly what does that argument prove; and to see that no fallacy creeps into the consequence. If the gentleman desires it, we shall give

a somewhat analogous argument, and draw the consequence clearly, in the terms of Carl Vogt: then
he will see what his own premises prove. In Europe and America, there are two parallel and inde-
pendent lines of horses coming down from remote
geological ages. These races of horses are of like
structure. Consequently, by evolution, they should
have had a common origin; just as our friend is referring the rudimentary organ in one type to a
common origin with the perfect organ in another,
simply because they are alike. Vogt affirms that
the facts are quite otherwise. The more one re-
cedes into geological ages, among the fossils of
these horses, the more does he find the races receding from one another; so that far from having diverged, on leaving a common ancestor in the past,
they have converged from different origins, and
merely assumed a common type in the present. I
quote these observations from the *Revue des Questions Scientifiques* for April, 1887, under the head,
Monophyletism; and they are useful not only for
gauging this rudimentary argument, but also for
suggesting a fitting reflection on Professor Huxley's
series of horses, and " their demonstrative evidence
of evolution" (No. 121). The whole argumentation
on the rudiments is either that of a false analogy,
or the fallacy of explaining the unknown and obscure by what is more so; pretending to prove and
not doing so. Still it passes current in science as
a proof good enough for Darwinism.

132. And not for Darwinism alone. It is only fair to state, that evolution of the wider, higher kind which introduces into its subject some evolving principle of progress from non-life to life, and from the lowest forms of cell-life to the highest complex organisms, goes two ways, and every way, no less than Darwinism; it goes up and down. The upward form is commonly called Evolution; the downward form is technically styled Catagenesis, or Degeneration. Hence they are found blended together, and a composite theory is formed of evolution upwards supplemented by downward phases of Catagenesis, thanks to the genius of Dohrn and Lankester. So that by combining a tree of ascent with a tree of descent, one with its roots below in spontaneous generation, according to Professor Haeckel, and thence shooting upwards to man, the other with its roots upwards in consciousness or sensibility, according to Professor Cope, and thence running down to the amphioxus and the ascidian, and " the polar tensions of chemism," a perfect spectacle is exhibited of the ascent of species, or descent, or both at the same time, like a mirage in the desert. Argumentation like this, in the anatomy of a system, makes it quite as interesting a specimen for study as the rudimentary teeth of the fœtal whale, or the legs with which the boa cannot run away, or the wings with which the ostrich cannot fly. Provisionally, we might subscribe so far to the Darwinian doctrine of rudi-

ments, as to consider this specimen of logic a
rudiment of some better state of philosophy and
thought, which may have graced the hapless scion
in its ancestry. Indeed, stretching our imagination
backward, to use Mr. Tyndall's happy phrase, "beyond the experimental boundaries," we may be
permitted to discern the secret of its birth. And
if so, we cannot fail to see that, child as it must be
with some ancestry or other, possibly at some point
of palæontology it had a noble sire.

We have finished with the idea of species, and
also of race. Having been discursive enough to
think now of closing, we shall add a few words on
the other Darwinian terms, and leave to the next
chapter all that remains of the criticism on Life,
Cells, and Evolution.

133. To finish the terms which occur in the present
question, we may mention that there are cross-
breeds between races, and, in spite of
the natural sterility so often spoken of,
there are cross-breeds between species.
The former, which are abundant in nature, are
called Mongrels; the latter, which are rare, go by
the name of Hybrids.

134. From this it appears that the test of natural
sterility is, as Dr. Romanes puts it, neither absolutely constant, nor constantly absolute. Never-
theless, as a test for the question of the possible
descent of species, it remains quite uncompromis-
ing, for all practical purposes. This is brought out

into bold relief by merely reporting the circumstances of successful hybridation, or the crossing of
species. In the first place, it is man's interference
that brings it about. Nature does not affect it;
though there are some native instances reported in
almost every order. M. Suchetet sums them up
in an exhaustive article on the subject, in the number for last July of the *Revue* mentioned before.
Secondly, hybrids are rare. As to their number in
the state of nature, considering that there are
143,000 species classified among animals by zoölogists, the facts of hybridism which are reported from
every quarter and through every channel, " might
be multiplied," says the same authority, " ten times
over without acquiring any importance in the question." So they are rare. Thirdly, they are sterile,
as is well known in the case of the mule, which is
the commonest and most uniformly successful experiment of hybridation. It is narrated that Arab
populations, which ought to be better acquainted
with the natural facts of horse, ass and mule than
any other people, have been thrown at times into
the depths of superstitious dread by the report circulating that a mule had been productive. Fourthly, however, hybrids have maintained themselves
for some time in a line of descent; and elaborate
experiments with animals and plants have shown
the phenomena that now come into play. It requires extraordinary care to maintain them at all.
In spite of the care, some of the offspring at each

generation revert to one or other of the specific
types which were crossed. Then they never resume the other type, as if they had rejected all par-
ticipation in a mixed specific nature. This is the
phenomenon of Reversion. Others, which do not
revert, throw all calculations about them into a
distressing state of confusion, by showing extraordinary and mutually divergent variations. This
is the phenomenon called Disordered Variation;
and it testifies to the irregularity which has been
put upon nature and which is continued through
man's interference. Fifthly, there is only one
hope, at length, of the hybrid progeny surviving,
and that is by absolute reversion to one or other
parental type; so that the case, as one of hybridation, is utterly extinguished. With mongrels (No.
133), on the contrary, the fertility is even increased
by the mixing of races; and, when a distinctive race
has been formed, individuals may still at any time
reproduce as native traits the special characters of
one or other parental race, which originally blended
to form this mongrel new one. Such a reproduction of ancestral traits is called Atavism.

135. The variability which gives rise to varieties
and races as the result of self-adjustment to the
environment, is not restricted to any single organ,
which adjusts itself in that manner. A proportionate adaptation takes place in other parts also. This
might be understood from what we said before on
acclimatization and naturalization (No. 77, 78).

Mr. Darwin has, however, applied a special term, that of the "law of correlation," to signify this proportionate adjustment.

136. Moreover, it is noticed that an organ, if long unused, can become reduced in weight and somewhat in size; just as by active use it Use and Nonmay become stimulated and developed. use. Observation has not shown that any unused organ becomes a mere rudimentary structure, as if forsooth non-use or disuse were a morbid affection supervening to destroy such local structure. And, as to the effects of use, one limit or condition is imposed by philosophy and common sense; it is that a thing or an organ must be, before it can be used and developed, or can use and develop itself. Yet Mr. Darwin introduces the element of non-use, as a sufficient reason for the disappearance of entire organs into that condition which he calls rudimentary, and which we criticized above. And he assigns the element of use, as an adequate suggestion why a local structure should begin to be; because under the touch and stimulus of environment, the general organism, that comes to want it, uses it and works it by minute degrees into existence. Here you have an instance of the marvelous simplicity which has charmed the minds of men, and won them over to Darwinism, with a magic more potent than logic.

137. But I am tempted to quote the master of anatomical science, George Cuvier, who in his

Comparative Anatomy speaks thus: " If any one should be bold enough to assert that a fish, by dint of standing up continually on dry land, would see its scales fall away and change into feathers; or that thus it would become a bird; or if any one tells us that a quadruped, by dint of squeezing itself through narrow ways and stretching itself out while walking, might change into a serpent, he would do nothing else but give proof of the most profound ignorance of anatomical science." And again, in his discourse on the Revolutions of the Globe, he instances the dog, which has accompanied man everywhere, has undergone all kinds of modifications, has in short been subjected to artificial selection in its fullest sense. Hence the different races of dogs differ in every conceivable way, as much, for instance, in their measurements as one to five; yet, he continues, " in spite of so many and such great differences, the relations of the bones remain the very same, and never does the form of the teeth differ in any point of consequence."

` 138. I do not venture to say whether science has improved on these very peremptory conclusions. They seem to be admitted now as much as in Cuvier's time. A walking animal, it is granted, cannot be descended from a climbing one. Vogt, in placing man among the primates, that is, among the apes, declares without hesitation that the lowest class of apes have passed the landmark (the common ancestor) from which, according to evolution, the different types of this family should have originated and diverged. And, in the most recent morphological speculations, I observe that it is pronounced a very unsafe principle " to make the apex of any one group in nature the base of the next," as if one type of anatomy could change into another. So that no amount of use can originate organs, and, still less, species; no matter what incantation of environment Mr. Darwin employs to invite them into existence, or what " law of permanent characterization" he invokes to fix a species, when once he has conjured it into being.

139. Finally, there remain three terms, the struggle for existence, the survival of the fittest, and natural selection. Mr. Darwin takes up an idea already familiar in the English school of political economy, and he reckons that every kind of animal and plant tends to multiply itself indefinitely, according to the ratio of a geometrical progression; just as Malthus had supposed in his theory of population as regards the human species. From this Mr. Darwin infers that every organic individual is put through a severe " struggle for existence," so that only the requisite few survive; just as Malthus had inferred that preventive and repressive measures should be employed to keep down human population. In point of fact, Mr. Wallace tells us that the sum total of animal and vegetable population remains almost stationary.

Struggle for Existence.

140. It will be useful however for both schools to observe that, in sound philosophy, no such supposition can be admitted, as that the economy of nature involves a perpetual and universal struggle, either physically among the brutes, or morally among men. Every problem which nature proposes she solves herself; and what is overdone or underdone in one direction, she makes compensation for in another. There is no need of interposing with such fictitious elements as a system of repression to control, or an imaginary struggle to interpret, what in the last resort is but the smoothly rolling economy of nature. A struggle for existence, if the words are taken to mean what they say, intimates a state of violence; and universal organic nature cannot be undergoing habitual violence. Hence the supposition implied in the meaning of the term, struggle for existence, must from a philosophical point of view be simply denied.

141. But as far as the term is taken to signify that endeavor on the part of organic beings to adjust themselves, as best they can, to their conditions of life, the idea is true and real, without being really a struggle. In this true sense, the play of such a factor, which is otherwise called acclimatization or naturalization, is only a small element in the problem, as to which individuals will actually survive. It expresses only, in the organic kingdoms generally, what we have already considered (No. 77–81), how nature tends to discriminate in

favor of the better qualified races, and less favorably with regard to the rest. Some of the less provided ones may even be eliminated; but it is all a matter of races, and never touches the question of species. It may exterminate a species; but it cannot create a new one, or transform an old one.

142. All that is only one factor in the problem. There are others besides. Perhaps it is not fair to ask many questions of an ill-digested hypothesis; or we should like to inquire **The Economy of Nature.** what possible application can this idea of a struggle for existence meet with universally in nature, when there are entire orders of beings evidently meant, not so much to exist themselves, as to keep others in existence by being eaten up continually; and therefore, as useful means to support the higher life of others, the individuals of these lower orders are multiplied in numbers beyond the grasp of ordinary mathematics. Without pretending to determine the rank of herrings, for instance, in the scale of being, we may consider the shoals of them that are provided for the whales and other big feasters of the deep. Their struggle to exist, if existence means fitness to do their work in life, must be to come forward and get themselves eaten up. Consider the spectacle on the banks of Newfoundland. All nature there, above and below water, is described as a system of gigantic depredation. Near the shore, the smaller classes of fish, such as can be netted in a pocket-handkerchief, are swallowed

up by the next larger; and so on out to sea, where
the full-grown cod-fish lies in wait to devour his
brethren and cousins in every degree, and literally
eats his way through a hundred miles into shore.
Upon the cod again, as well as upon the smaller
fry which he pursues, rushes the greedy shark, the
" bottle-nose," a small species of whale, the " pot-
head," the porpoise, and all the big marauders of
the ocean.   Or, again, consider those other orders,
the microbes, which are not meant principally to
be eaten, but rather to eat up and eat out of organic existence all bodies of higher complex structure, as soon as dead.   These, if not quickly disintegrated, would keep large quantities of matter
locked up, so to say, and lying idle in a dead state,
instead of being out and free in physical circulation.
Nature sets on them the microbe, multiplying any
single one of these microscopic animals at the rate
of a million millions in a short season.   Now what
does the struggle for existence mean among them ?
To come forward and eat ?   Yes; but then they
die, countless billions of them, with the thing they
eat; and nothing alive remains of either.   The elements of all are circulating freely again, in air or
water, to provide for the general life of other species in creation, which need those same elements.
It is not precisely a struggle here for the individual's own life.   It is a work going on under a higher
principle in nature, which is providing for one order by means of another; just as in the human

body there are organs generating secretions, which they themselves do not need and cannot use, but which other parts of the system require. So Dr. Foster remarks of the function which produces glycogen: " Obviously the organ makes this not for itself, but for other parts of the body; it labors to produce, but they make use of, the precious material, which thus becomes a bond of union between the two." Is anything more desired to show a design over things, than that one should labor, and, if you like, " struggle " and perhaps die in the work, while another reaps the fruit; and that all the while the economy of nature should roll on indefectibly and smoothly? Mr. Darwin does not understand things thus: he means a struggle for the individual's own existence. In that case, whether you take it as struggling or as simply living for the individual's own solitary good, there is but a limited play of such individualism in the divinely contrived course of nature.

143. Equally restricted is the meaning which underlies the other phrase of Mr. Darwin's, "the survival of the fittest." This survival should mean for his purpose, as it cer- **Survival of the Fittest.** tainly does mean for the evolution which moves on parallel lines with his, the origin of higher species in the world, a general trending upwards of the organic orders, the lower species which survive becoming gradually higher. In this way, the lower orders which are meant to be the food of the higher,

the simpler and plainer which evidently sustain the more complex and specialized, would pass out of the stage of species more generally eatable and take rank among the higher and higher, which are more generally the eaters; and the lower ranks would be vacaied. But we find a pronounced evolutionist declaring that such an application or inference of evolution will never suit the nutritive arrangements of the universe. M. Gaudry says in his Primary Fossils: "There would be more superior animals than inferior, more eaters than beasts to be eaten; the harmony of the universe would long ago have been broken." So he requires that some of the lower orders should not have advanced in the upward march of evolution; they must benignly have remained content with their humble lot, of remaining edible and acceptable to their betters.— Does it not look as if some mild symptoms of catagenesis, or evolution downwards, broke in upon us here, when we did not expect it? The gentle infirmity makes the system even more engaging, particularly when with such simplicity it serves the interest of truth.

144. The true statement of this factor, the survival of the fittest, will be as follows. It signifies the process of adaptation to environment; and here those varieties in a species have the best chance to live and thrive which are acclimatized, whatever direction their self-adjustment may take; either towards the absolutely better, or the absolutely

worse. For, to illustrate lower things by higher, it is thus that an inferior race of men can survive in an African marsh, where the flower of Europe would die; and, in fact, the degenerate races of Africa do survive as the fittest where they are, because, in the struggle of centuries, endeavoring to escape the evils of tyranny and slavery, they have fled into the midst of poisonous miasmas; and, degenerating to a degree of conformity with the worst conditions of life, they have grown acclimatized there and survive. This just illustrates what Mr. Wallace notes of Darwin's theory, that the minute variations contemplated in it are really of any kind, and in any direction. And the survival of the fittest means simply the course of nature, whereby those survive who are fit to survive, or it may be, are the fittest, but are not necessarily the best. Therefore the theory has nothing to do with the upward development even of races; much less of species.

145. Now, if the struggle for existence comes to designate only the economy of nature in the organic kingdoms, and similarly the survival of the fittest denotes the course of nature, what remains to be covered by the **Natural Selection.** final term, "natural selection," which is meant to convey all the rest ? Just so, it conveys all the rest; and that, as we see, is little enough. The rest tells us little that is new. Besides, we have seen that this complex term is logically a misnomer (No. 106), as

compounded of simple terms which are incompatible: if it is a selection that is really meant, it must be artificial, not natural; if it is natural, as opposed to artificial, then it is no selection. There is a verbal equivocation in using the same term for two different ideas. But if the conception was honestly intended, then it rests upon a mistaken and false analogy, as we hinted at in the same place. So that to conclude with the words of M. Jean d'Estienne, in the number for January, 1889, of the *Revue* which I have quoted before: " Reduced to its most simple expression, stripped of all seduction of style, and of all logical artifice, Darwinism comes down to a very small affair. It is a system of hypotheses ingeniously applied to the justification of a first hypothesis."

We proceed next to the origin of life and its progress on this globe. This will lead us to examine the Cell, and the march of Evolution.

BIOLOGY,—(*Continued.*)

CHAPTER IV.

CELLS; OR, EVOLUTION.

146. In the seventeenth century, an Italian naturalist of eminence, by name Redi, undertook to sift the question of spontaneous generation, which we otherwise call organic evolution (No. 102). It was a question older than Aristotle, whether life could spring from non-life, living things from non-living. They seemed to do so, as every one has thought he saw for himself in decaying organic matter, in meat, cheese, and the like. Where does the life come from, when it suddenly appears there, unless it starts up of itself from inanimate matter, and from the mere chemical elements? Now this looks like the evolution of organisms out of the inorganic. Hence the name we give to it, organic evolution. But the system of philosophy then prevalent, that which is called the Aristotelian system, was adverse to such a view. Even in the absence of ocular evidence to support its position, it affirmed, *Omne vivum ex vivo*, "The living comes only from the living." It preferred to fall back upon the active energy of the sun, holding that to be a kind of universal cause equivalent to organic parentage, rather than admit that life could spring from inorganic matter without

any parental agency.   It is a little singular, by the
way, that the diametrically opposite school to-day
of scientific materialism has recourse to the same
"potential energy of the sun," for precisely the
opposite conclusion.   .

147. The experiments which Redi performed
were not unworthy of science, for the epoch at
which he lived.   By exposing meat in a couple of
jars, leaving one open to the insects of the air, and
closing the other, he found as he had surmised that
no life sprang into being where insects were excluded, and the usual abundance of minute animal-
cular life appeared, wherever they did enjoy free
access.   Some exceptions were taken to certain features of his experiment, but upon his modifying it,
the same results were obtained; and science was
satisfied.   The microscope had not as yet enabled
observers to detect the vast world of life, which remained still invisible in decaying organisms.

148. When the microscope came into use, and
especially when the principle of achromatism was
applied to it by Professor Ehrenberg, a
The Micro-   world of revelations broke upon the
scope and
Animalcules.   astonished eyes of science.   Living
things were observed breeding with extreme rapidity in water that was poured over, or
"infused upon," dead organic matter; and the Professor called them *infusoria*, or infusory animalcules.
Where did these come from?   Surely, spontaneous
generation was now reinstated and proved.

149. The animalcules thus discovered were, in course of time, found to comprise a great variety of minute living beings, which had nothing in common except their microscopic minuteness. Plants as well as animals, mollusks, crustaceans, insects, and worms, larvæ and perfect forms, were all found to have been massed indiscriminately under one vague term, animalcules. Consequently, that term is now taken in no specific sense; but may be used to signify all infinitesimal organic beings, of any size, from the hundredth part of an inch to a minuteness which the glass can scarcely distinguish, though magnifying its object thousands of times.

150. Among them there is one class which is of supreme interest for the present question. We are examining here the development and origin of life, or the living organism which Darwin endeavors to transform from one species across to another; which evolution, taken in a wider sense than Darwinism, endeavors to transform from the lowest and the lower species up to the higher and the highest; and which on the same principle should be found in the lowest and simplest species only one remove, if at all removed, from inorganic elements, from mere chemical and physical forces. Now here among the animalcules is found a class of living things which are composed of single cells. The term cell designates the smallest integral portion of matter which can exhibit life, can receive it, or can communicate

The Uni-cellular Animalcule.

it to another. It is the physical unit. And life, it is now found, may be entire in little organisms which consist each of only one cell, and are called "unicellular." In the great organisms too, consisting of many cells, and called "multicellular," it is still the same physical unit of organic matter that the organism extracts from the pabulum, transforms into fitting material by digestion, informs with life, and which, as a new living cell, it adds to its structure in skin, bone, cartilage, muscle, etc. In every case therefore, the cell is conceived to be the last physical unit, wherein life can be found, and in less than which life cannot be.

151. Inside the wall or envelope of the cell, and sometimes constituting the whole of it, is found an element which seems indispensable. It

Protoplasm. is a liquid substance of slimy consistency; and, to follow the description of Gordon Salamon in his recent lectures on Yeast, delivered before the Society of Arts, it is "endowed with specific organization, and is capable of exhibiting motion." This substance has been called protoplasm. Sometimes, as in the early stages of the life-history of certain organisms like the slime fungi, the protoplasm is not contained in any cell envelope whatever; yet it can express its vitality in terms of motion and constructive increase, that is, it can move, grow, and multiply. Life is unknown without the presence of this protoplasm, which is to be found in every organic living cell, whether animal or vegetable. Accordingly, it has been treated of by Professor Huxley as "the physical basis of life."

152. Here the interesting question arose, whether a chemical compound of this kind, understood to be composed like all organic matter of oxygen, hydrogen, nitrogen, carbon, re- ^{Its} Chemistry. quired parental generation to produce it. True, it was alive. But in its simplest forms it is not embarrassed with any organs or local structures for special work; it has no eyes, ears, mouth, stomach, feet, head, or tail. Is not that the condition of chemical elements,—they are unorganized and inorganic? What prevents the unorganized protoplasm from being generated by the inorganic elements; the more so, as these are now compounded in the laboratory to degrees of complexity more than sufficient, it would seem, to equal this rude fabric of cellular protoplasm?

153. Unorganized and inorganic! You see room here for a brilliant equivocation, one indeed that went some way for a while towards electrifying an enlightened age. Imagine the "inorganic" chemical compound, which discharges no functions, because forsooth it is not alive, becoming in the chemist's hands live protoplasm, which is "unorganized" also, though forsooth it discharges all functions, and it moves and eats and digests and sleeps, as higher structures do, though they do all these things through a number of local organs.

You have only to consult Professor Leidy's monograph on the Fresh-Water Rhizopods, to see what
these little live cells can do and how they do it.
But, laying no stress on this minor point of complete life being there, and regarding only the ground
for equivocation, we need not wonder at Professor
Huxley, in his essay on the Physical Basis of Life,
throwing out the phrase, "protoplasm dead or alive,"
as if there were no important antithesis conveyed in
the term, "dead or alive." Alive, dead, and never
being either one or the other, have usually been
reckoned three different ideas; but not so, apparently, in this chemistry which sinks all in one
phrase, dead and alive!

154. So protoplasm seemed to command the
position between life and non-life. There was this
little cell, consisting chiefly, if not entirely, of protoplasm, exhibiting a positive vitality, and a most negative organic
simplicity. And chemistry, feeling easily
sure of the simplicity, thought that now at last it had
the vitality also. A great future was dawning on
science; and the past history of the world was being
unrolled. It was by this stage that life must have
originally walked upon the globe. Who knows!
We were thrown back with Mr. Tyndall to "the
possible play of molecules in a cooling planet."
Here protoplasm betrayed that first playing of life.
And the term protoplasm was devised to express
the first plastic development thereof; for the Greek

**The Border-
ground of
Life.**

word, *protos*, signifies " first." At once it was interesting to see how at this plastic stage in the possi-
bilities of scientific development, a whole series of
primary elements sprang up in biology as by a kindred instinct of spontaneous generation. There
were the orders of protists and protophytes and pro-
tozoa, etc. And what added to the completeness of
the view, and the absolute establishment of organic
evolution, was that, at this stage of life, animal and
plant are indistinguishable. They were confounded
by Ehrenberg; and, if one evolved by spontaneous
generation, so could the other as well; yes, and we
may add, they could just as easily evolve from one
another besides.

155. All this is now a romance of the past, be-
longing to the dim border-land where
dreams and imagination love to dwell.
As Gordon Salamon remarked, " Even
protoplasm has a specific organization
of its own." And, equivocation evaporating from
the moist product of imagination, the light of science waxes strong, and the chemistry of the question is precipitated to its own sedimentary level:
vitality rises to its proper sphere; and, between the
two, spontaneous generation, that again seemed to
be, has once more ceased to be.

156. A negative and a positive manner of proving
this have been successfully adopted; and the results are universally accepted. That was a negative way which was followed by Schwann, Van Ben-

eden, Pasteur, Tyndall. It consisted in precluding all possibility of live germs penetrating into a given medium: the medium contained some dead organic matter, in which such germs or embryos of animal-cular, bacterial or microbic life, as you choose to call it, are known to grow: then it was kept under strict examination, to observe whether, these conditions being rigidly kept, any life appeared in the dead material. Notice then the conditions of the problem: a solution, called proteinaceous, is provided, one capable of the highest putrescence, and therefore prime material for putrefactive germs to settle on and develop in; but the solution is absolutely sterilized, that is, cleansed of every particle which can possibly be a germ of life; and it is placed in an optically pure, or absolutely calcined, air. Now, the conclusion ascertained is this: while such conditions are maintained, no matter what length of time may be suffered to elapse, that putrescible fluid will remain absolutely without trace of decay. There is no putrefactive life in it, no microbes, no bacteria, simply because there is no antecedent parental life there to produce them. Such is the exclusive, the negative way of proving that life can come only from life. *Omne vivum ex vivo.*

157. There remains the positive system, which will take up the direct study of these animalcules, and will tell us how each has come into being, upon its being found to develop in any given medium.

Here a double line of investigation is open. One
fixes upon a given organism, in a medium
wherein that organism alone is "cultivated." The other takes them indiscriminately as they appear in natural
conditions, and catches nature, as some one 'has
said, " off her guard." In the first line of investigation, a fitting pabulum is prepared, that is, an extract of meat carefully filtered from other kinds of
life, or an extract of fruit; and into this is admitted
a germ of the solitary kind of microbe which comes
up for examination. In this pabulum it develops
and multiplies. Now you have only to watch it;
and that is done to the extreme degree of scientific
solicitude by what has been called the unbroken-
watching system; a couple of competent observers
relieving one another, so that the object under in-
spection is never lost to the eye for a moment, not
even during the space of many days.

158. As an example of special cultivation we may
mention among others Dr. Koch's treatment, two
or three years ago, of the germ which he considered
to be the cause of pulmonary comsumption, for it
was uniformly found in the epithelium of diseased
lungs. To test whether this was the specific cause
of that disease, he took a germ from a diseased lung,
and placed it in a medium where it would thrive and
multiply, but from which every other species of microbe had been carefully excluded. Then it was
necessary to go on cultivating this microscopic

Positive Argument.
1. By Culture.

thing, letting it grow, while modifying its surroundings in various ways, to see what would become of
it, and whether perhaps it would turn into something else.   Dr. Koch cultivated his stock of
phthisis microbes so far that they lost nearly all
their virus, and were brought to the very verge of
sterility, and still the stock remained unchanged in
species.   Identifying it thus as a distinct species,
he wanted to find out further, whether the identical
disease of consumption could be communicated by
it to a healthy lung.   Now the cat is understood to
be particularly exempt from the attacks of this disease.   So he communicated the microbe to a cat.
The animal became consumptive.   And thus he
proved the disease to be contagious.

159. In the case before us, the object is merely
to see whence it is that the given organism under
view takes its rise; what is its life-history, or life-
cycle; and whether it has the character of a species remaining unchanged, as like always produces
like.   To cite the results of the Rev. Dr. Dallinger's observations on the *bacterium termo*, which he
identifies as the exciting cause of all putrefaction,
just as the yeast-plant is the exciting cause of what
is commonly called fermentation, " these organisms, lowly and little as they are, arise in fertilized
parental products.   There is no more caprice in
their origin than in that of a crustacean or a bird."
Referring to the negative proof given before against
spontaneous generation, he says: " By experiment

it is established that living forms do not now arise
in dead matter. And, by the study of the forms
themselves, it is proved that, like all the more complex forms above them, they arise in parental products. The law is as ever, only that which is living can give origin to that which lives," *omne vivum
ex vivo.* For the particulars of his observations I
refer you to his lecture, " Researches on the origin
and life-histories of the least and lowest living
things," in the scientific journal, *Nature*, October
23 and 30, 1884.

160. This then is the first way, that of cultivating a special germ. Another way is that 2. By Obser-
of following these organisms while they vation in
are out in their own line of life, and Nature.
watching them in their natural conditions. In
these conditions, a number of forms are found together, as Ehrenberg observed; or they work con-
secutively on a given material, as Dr. Dallinger describes. I can but refer you again to the instruct-
ive account which this latter observer made the
subject of his address, on retiring from the presidency of the Royal Microscopical Society, last
spring. It may be found in the *Scientific American
Supplement*, April 28, 1888. His subject was restricted to those organisms which do that special
work of fermentation, called putrefaction or decay.
But his conclusions cover the whole ground before
us. There are putrefactive organisms resembling
in form those other microbes which are parasitic

or pathogenic, like the phthisis microbe mentioned
above; that is to say, which are capable of developing disease.   However alike in form, these sets
of organisms are different in function; and, no less
than the great organisms in nature, they are different in species.   It appears in short that there is a
whole series of microscopic animals, which are related indeed, and which are altogether alike to the
eye, or are more or less so; but they are in many
respects greatly unlike, embryologically and physiologically.   " This is a region of life in which we
touch, as it were, the very margin of living things.
If nature were capricious anywhere, we might expect
to find her so here.   If her methods were in a slov-
enly or only half-determined condition, we might
expect to find them here.   But it is not so.   Through
years of the closest observation it will be seen that
the vegetative and vital processes generally of the
very simplest and lowliest life-forms are as much
directed and controlled by immutable laws as the
most complex and elevated."   This then negatively
and positively settles the question of spontaneous
generation, which we have otherwise called organic
evolution.

161. That was one feat of science to discover
such beings as consisted only of a single
Multicellular Organisms. cell, whether in the plant or animal order.
It was another achievement to unveil
the great fact that every organic being, no matter
how complex and perfect and individual in its unity

and totality, consists only of cells and nothing else.
Skin, bone, cartilage, nerve, are very different to
the eye and touch; but they are all composed of
the self-same thing, cells. The cells, it appears,
are the same throughout in their origin and development, till they become adult and perfect, when they
are all found specialized so as to discharge distinct
functions. In view of these distinct functions, the
anatomical form of the cells has become differentiated, some taking one form, others another, cylin-
drical, hexagonal, polygonal, conical, pyramidal, len-
ticular, according to the place and work before them;
whether it be that of the muscular tissue, cartilag-
inous, osseous, fibrous, vascular, or the like. Thus
the entire system, in its totality and individual unity,
is built up of many cells, and only cells: these go
to form all its organs.

162. Each individual cell is first young and then
old. In the same organism while some cells are
only beginning their development, while others are
carrying on theirs, another set are in full decrepitude,
and a number are being disintegrated as effete. In
the young state of cells, to which stage is to be referred also the earliest condition of an embryo, that
is, the germ of another complete organism, it is not
possible to distinguish the growing, moving matter
which is to evolve an oak, from that which is the
germ of a vertebrate animal. Nor, in the building
up of the same organism, can any difference be discerned between the germinal matter of the lowest,

epithelial scale of man's organism, and that from
which the nerve cells of his brain are to be evolved.
Hence you see that since the first germ from which
any organism springs, no matter how humble or
how elevated, whether it is to be of one cell or of
many cells, simple or highly complex, is just like the
first germ from which any other organism, whether
plant or animal, arises, it is not strange if, in their
primary stages, all organisms are indistinguishable
from one another. Yet strange inferences have
been drawn from this very plain fact, as we shall see
later on (No. 191).

163. You see too how a new form can be given
to the old axiom; and, instead of putting it *omne
vivum ex vivo*, it may be changed into
omne vivum ex cellula. Biology, in the
last analysis, becomes now a study of
cells. And, in its highest synthesis, it remains
largely a study of cells; for the living being consists
in its entirety only of these same elements, diversely
modified for diverse functions, and so constituting
various organs. The description then of all life, as
viewed from this material side of the element which
goes to compose it; which is assumed from the
pabulum or food, is developed for a special work
elsewhere, and then is worn out as effete; may be
comprised in still another shape of the same formula; which now becomes, *omnis cellula ex cellula*,
" cell from cell."

164. The embryo which begins even the highest

*Omne vivum
ex cellula.*

organic structure is only one of those cellular specks of protoplasm. Then it develops into a multicellular mass, without having as Embryonic Development. yet any distinct local structures, or organs, as they are called. Becoming the subject of further changes, which tend to the production of a complex structure, it now begins to consist of parts different, but mutually dependent. At the respective stages of this development, certain generic likenesses obtain between it and other embryos — rather negative likenesses than positive, becoming less as the positive qualities and organs of each come out more; until all is lost in that specific and unique identity being assumed, which is unmistakable. With regard to these generic or negative likenesses, Mr. Herbert Spencer observes in his Principles of Biology: " The resemblances which hold together great groups of embryos in their early stages, and which hold together smaller and smaller groups in their later stages, are not special and exact, but general or approximate; and in some cases the conformity to this law is very imperfect."

165. Just one remark here. People go to much pains in finding out points of likeness The Cell's " Glorious Mission." between species, to prove their transformation. But here is the most plausible fact of all, that every organism in every species is made up of cells, and every cell of itself is like every cell in any other being, save only in the form

which it assumes according to its place in the structure. If matter is everything, as it is for the materialist, why should not anything change into anything, since there is only the organic cell everywhere—if, I say, matter is everything. Nor is it
surprising that Professor Haeckel, an industrious
disciple of materialism, should give the cell "a
glorious mission;" albeit some of his fellow-dis-
ciples, like Edmond Perrier of the same school,
rather enjoy a laugh at his enthusiasm. For contemplating Haeckel's moneron or formless cell, and
the genealogical tree of *phylogenesis* (No.191 below),
sprung from that cell and from his enthusiasm into
every order of species, the materialist critic sneers
at materialistic ingenuity,—" as if a certain living
being had received the glorious mission of conduct-
ing life to its most elevated form, all the way up the
ladder of animal species,without once stumbling on
any one of those other forms, which were destined
to stay down, but with which it just acknowledges
common ancestors and (collateral) cousins!"

166. Let us examine what truth there may be, like
a germinal cell in history, to which such
a flourishing tree of theory may trace its
genealogy. Throw yourself back to the
time when, according to the nebular
hypothesis, this planet of ours, like so
many others, was still in a nebulous, fiery state, one
of intense light and heat. There was no question
of cells developing as yet, nor of any cells existing

The Nebular
Hypothesis.
First and
Second Days
of Moses.

in such a furnace. Ages rolled on and this fiery
cloud gradually cooled. Rocks and metals, which
were thus far in a vaporous condition, came into
contact with the cold of surrounding space. "Void
and empty" as the planet itself was, it rolled in the
midst of space still more so. The vapors of rocks
and metals thus began to liquefy, and, like condens-
ing clouds, fell upon the earth in showers of molten
metal; and, if they rose again in vapors, they settled
again in a liquid state; until, by the continued loss
of heat, a thin crust of solidified metal and rock
formed on the surface of the burning mass. Radi-
ating heat as it continued to do into space, the
whole earth grew cooler every day. And the time
came when even the aqueous vapors of the planet
began to settle from their condition of steam into
that of liquid and seething water. And the rolling waters began to form sedimentary deposits,
from the wearing of the rocks, as soon as they be
gan to roll. Continuous rains now prevailed, maintaining a thick darkness, " a cloud as a garment, and
a mist as swaddling-bands," over the face of the
waters, all above and around upon that boundless
ocean which knew no shores, and had none; until
the constant upheavals of the thin crust of the earth
gave it "bounds and a bar and doors," and there it
began to "break its swelling waves." The continuous rains from an unbroken belt of vapors
gradually ceasing, the vaporous clouds were broken.
The waters of the clouds above were separated from

the waters of the ocean below, and suspended in a permanent atmosphere of lighter gases, which Moses calls an " expanse" or " firmament," dividing the waters beneath from the waters above.

167. The crust of the earth heaving upwards in some places, sank in other parts; and the waters gathered together here, in seas, while the dry land appeared there.   Those parts

Third Day.

which were never to be the home of man sank, never to rise more.   Mr. John Murray informs the world, as the last results of the " Challenger" explorations (No. 67 above), that the abysmal regions of the Atlantic are unique: they have never been elevated, never a continent of dry land.   " The result of many lines of investigation seem to show that in the abysmal regions we have the most permanent areas of the earth's surface." I refer to his account, as reported in *Nature*, October 15, 1885.   And Sir J. W. Dawson, of Canada, informed the British Association in September of the following year, that " the history of ocean and continent is an example of progressive design, quite as much as that of living things."

168. There was now land and sea.   There was heat—an excess of it, even in the waters.   And there was light.   The sun was not as yet; and the light was a phosphorescent or nebulous one, coming from the molten, central mass, which, having thrown off the earth already, had yet to throw off other planets, and then to be condensed into a sun.

None of those phenomena therefore which depend upon the special solar energy, could as yet appear upon the earth, when, beginning a new phase of activity, it now put forth the simplest forms of life, or, as Moses says, it "brought forth the green herb, and such as may seed." Here is the cell and cellular life, for the first time. It is noteworthy that the lowest kind of plant, and the lowest kind of animal, protophyte and protozoon as they are called, do not require the sun's special activity for their development. More than that: there are herbaceous trees, rich in pith, which, unlike the forest trees that grow by concentric rings in the revolving seasons, require no more conditions for their life than then prevailed, a warm soil, great humidity, an atmosphere saturated with carbonic acid gas. At this age then, besides the low form of vegetation which spread over the marshy land, there came upon the earth a carboniferous period, called by geologists the paradise of vegetation. Now was laid up a great part of that carbonized fibre, with which man would yet make himself comfortable, and make other resources of the earth useful, by mining it as coal,—a permanent reservoir of so much heat and activity once lavishly spent upon the globe in preparation for his coming.

169. The great nebulous mass in the centre, which had thrown off the earth as a ring or a planet to wander thenceforth in an **Fourth Day.** orbit of its own, threw off its last contribution to

the bejewelling of our system with the revolving gems of light; and then itself condensed into the nucleus which now we call the Sun.   That and the earth's own satellite, the Moon, stood henceforth in the firmament of heaven, both of them to be inextricably bound up with all the experiences of man's physical and intellectual life, of his moral and social history; determining the phenomena of day and night, of seasons and years; the one tinging with its golden rays the heyday of his prosperity and his glory, the other silvering with its pensive sheen the silent solitude of his path here below; both acting as tutors by their coming and their going, their radiance and their clouded brows, to his wisdom and his fancy, to his choicest prose and his deepest verse; verily, as Moses says, "two lights to rule the day and the night, and to divide the light and the darkness.   And God saw that it was good;" and our reason sees it too, and praises Him.   It praises Him as our first parents did, when, according to the poet, " forth they came to open sight of day-spring, and the sun, who, scarce uprisen, with wheels yet hovering o'er the ocean brim, shot parallel to the earth his dewy ray." They thus began, says Milton:

> These are Thy glorious works, Parent of good,
> Almighty!   Thine this universal frame,
> Thus wondrous fair; Thyself how wondrous then!
> Thou Sun, of this great world both eye and soul,
> Acknowledge Him thy greater; sound His praise

In thy eternal course, both when thou climb'st,
And when high noon hast gain'd, and when thou fall'st.
Moon, that now meet'st the orient sun, now fliest,
. . . . . . Let your ceaseless change
Vary to our great Maker still new praise.

170. Under the genial action of the solar energy, an expansion of organic types, to adopt Professor Dana's phrase, took place upon the globe; and there were found at once, be- **Fifth Day.** sides the lowest orders of life, also the highest orders, but represented only in their lowest species. Such as these were sufficiently provided for in the improving conditions of life. Many species already existing reached their stage of greatest prosperity, in the evolving conditions which suited them best; they passed that stage, and declined, either to die out entirely, or to survive in a few families. The tribes that culminated so are those of the crinoids, brachiopods, trilobites, ganoid fishes, amphibians, true reptiles, mollusks. Many other tribes have their era of culmination now, as gasteropods, birds, higher insects, teliost fishes. Brute mammals were to reach their climax in the Champlain period of the quaternary. And all creation, as we shall see, was to culminate in man, who never rose upon an inferior order of his own kind, and is never to be superseded or decline. His culmination is elsewhere; and the fortunes of the earth culminate in him.

171. When the sun then by his beaming presence, and the seasons which he controlled, had made the

earth more and more suited for the development of
higher types, then, to use the historical language of
Moses, was the proper age of the lower animals,
" those that swarm in the waters, and the creeping
and flying species of the land." They were all
over the face of creation, and they represented in
comprehensive groups the main types of nature.
This is the fifth day of Moses. In the plant and
animal kingdoms alike, the sub-kingdoms are all
present; the grand divisions are defined. The
specimens, which are representative of such divisions, look perhaps somewhat as if they were a common type of many other different forms, which
are more specialized and yet to come. But that
is always the case with things less perfect: they
are more common or general. A generic likeness
agrees with all its specific forms in a negative
way, inasmuch as it does not exhibit the perfec-
tions which the specialized forms display.

172. In the more imperfect conditions of life, the
more common type is the precursor of its
Sixth Day :
First Part. betters in better conditions which now
follow. The mammals begin to be; or,
in the terms of Moses, " the beasts of the earth according to their kinds are brought forth, and cattle,
and everything that creepeth (or prowleth) on the
earth." It is to be observed, however, that precursor does not mean ancestor. And the progress of species does not mean the descent of
species.

173. Nay, the progress of species does not even mean, in any uniform sense, that the earliest species under a type were necessarily the lowest, **Progress not** to be followed always by a higher. **a Descent of** Echinoderms, or the highest type of **Species.** radiates, were represented by species called cystids and crinids, long before the inferior type of polyps existed. The highest group of cryptogams, the ground pines, were a prevailing form of terrestrial vegetation long before there were mosses. There were huge crocodilians in the world, long before there were limbless snakes.

174. To do justice to the good taste of Mr. Darwin, as also of Messrs. Haeckel and Vogt, in admitting that, after all, at least the principal species of animals must have sprung from different origins, we must remember that the systems of animal structure classified as the zoöphyte, the radiate, the molluscan, the articulate and the vertebrate, are irreducible to one another: they are not on the same plan. "Can it be said," argues Flourens, "that there is only one form of nervous system? Is the nervous system of the zoöphyte the same as that of the mollusk, that of the mollusk the same as that of the articulate animal, etc.? If not, how can they be of one type?" And, carrying back the same analysis to the embryos of these classes, Muller says: "The human embryo never resembles a radiate, an insect, a mollusk, a worm. The plan of formation in these animals is altogether different

from that of the vertebrates." "No," says Milne-Edwards on the same topic, "a mollusk or an annelid is not the embryo of a mammal arrested at a certain stage of its development just as (among vertebrates) a mammal is not a fish perfected. Each animal carries with itself from the very beginning the principle of its own specific individuality; and it is a condition of its very existence that its organism evolve on the structural plan proper to its own species." Such being the case, one type of structure being irreducible to another, and incapable of evolving out of another, Mr. Darwin postulates four of five origins for species; Haeckel and Vogt likewise require several.

175. And what does geology record? Distinct and abruptly divided origins for the main classes of species, or sub-kingdoms; and, also for subordinate types under a main one, neither an origin nicely graduating from the more general one, nor even an origin subsequent to it at all, or contemporaneous, but in many cases preceding it; as in the echinoderms, ground pines, crocodilians just mentioned (No. 173).

176. As to the main divisions, the first vertebrates, fishes, start off suddenly in the upper Silurian age. This probably corresponds *No Transitional Types.* with the fifth day of the Mosaic account or cosmogony. No trace of links, or of "a finely graduated chain," to use Mr. Darwin's term, connecting the vertebrate fish with mollusk

or articulate, has been found. There are gaps everywhere. The ascidian, a mollusk, and the amphioxus,
a fish, have been dragged in to represent a transition
or bridge between mollusk and fish. But with such
unfortunate links as these, the same difficulty arises
as with all intermediate species interposed between
two widely separated ones. A criticism on a single
case like this may suit for all. I say then, it is like
bridging over a sea without a bridge: like finding
new islands everywhere, which require as many new
bridges. For species, to use Mr. Herbert Spencer's
idea, are only new clusters thrown out into space.
Or, like the shoots of a tree, the more there are,
the more the apices or points that shoot out towards
the sky. Yet the modern mind argues thus: "In
1860," says Professor Cope, "there were 250 species
of extinct mammals known: there are now something near 2000. I have found many myself."
That is to say, he must also have found many new
gaps between the newly-found species. Straight-
way, he goes on to infer by a "practical law of inference" that the gaps will all be filled up, or bridged
over for evolution to cross, because he has found
so many new species. But we may ask, how many
species new or old, all divided uncompromisingly
from one another, will make a connection or transition between any two? However, this is the Professor who requires only the rule, the measure, the
line, to determine what a species is or is not (No.
126). Another, who seems to understand something

of what a species is, argues thus in behalf of evolution, against the *Etudes Religieuses,* etc., of last May:
" The drag-nets of scientific explorers discover such
a variety of zoölogical forms, that it is often impossible to apply to them the best systems of classification so far adopted: types of transition abound:
between groups heretofore considered sharply separated,   there   are   found   intermediary   groups;
and often one species differs from neighboring
species only by imperceptible shades." All this
reminds us, in the first place, of Dr. Romanes' strict-
ure on the "small and trivial differences of form"
which distinguish species; and the vicious circle
which he finds in such argumentation (No. 115).
And, secondly, supposing that they are, as most
of them are not, true species, then what is it
that is really meant by these "types of transition
abounding"? Only this: that, between species
known before, new ones are found, not known before. But, if so, the difficulties are multiplied, and
the effort of engineering to bridge over the new
chasms in all directions is only intensified.

177. It is quite superfluous, at this date in the
history of science, to dwell upon the absence of
transitional types, such as might supply evolution
with a passage or line of march, from any one
species to any other. Fossil reptiles, the group of
whales, the tortoises and turtles, the frogs and
toads, these and many others, extant as well as extinct, came in without saying how they came. The

series of horses came in. But far from being transitional, one changing into another, as Professor
Huxley contended (No. 121, 131), "they differ from
each other in a greater degree," says Professor
Owen as quoted by Mivart in his Genesis of Species, "than do the horse, zebra and ass," which are
distinct physiological species. There came in those
extinct forms, birds with teeth in their jaws, and
with long tails, biped reptiles with the hollow bones
and some other characteristics of birds. I will not
distress you with the big names of archæopteryx,
odontornis, etc. All these are held to be transitional—a very obscure term covering a very gratuitous idea. Let us use a distinct term, and say, they
are gradational, in the sense explained before, that
"nature is continuous" or gradational, in her species and races alike (No. 121). They only exhibit
gradations, for in fact many of them subsist along
with the very species which they are supposed to
join by passing over from one to the other. J.
Barrande studied with conscientious care 350 forms
of the trilobites of Bohemia: only ten of these
varied at all in the whole depth of the strata which
contained them: the variations did not interfere
with their specific characters; and, instead of becoming more pronounced in time, they left the
species at the end of the record as it was at the beginning. Hilgendorf found twenty types of the mul-
tiform *planorbis;* instead of their evolving from one
another, he found them subsisting at the same

time. In short, these gradational shades have long
been acknowledged in the well-known circumstance
that so many different systems of classification are
devised, to describe nature systematically, if that
be possible. But the fact is, nature is like a net-
work, and nowhere exhibits a line such as evolution demands for its march upwards.

178. After these general strictures on the value
of "transitional" forms, let us take up the particular case of the mollusk, called ascidian,
and the very low form of fish, called
amphioxus, which have been put for-
ward to fill the gap between the lower orders and
the vertebrate fishes. For otherwise these latter,
to the scandal of evolution, come in at the stage
of the upper Silurian, without any transitional
form to usher them in. As to the ascidian, Verrill
has observed, after a thorough study of this molluscan tribe, that its alleged relation to the vertebrates
is without the slightest foundation in its structure.
On the other hand, the vertebrate, amphioxus, being a fish, agrees with the fishes; having no brain,
it seems to agree with the mollusks. Hence the
argument is implied that, after being originally
a brainless mollusk, it made the passage over to
the fishes by thinking itself into a vertebral col-
umn; and it left the mollusks on one side and
got among the fishes on the other. I think we
may leave this amphioxus to the tender mercies of
Catagenesis or Degeneration (No. 132), which has

The Ascidian and the Amphioxus.

despatched it sufficiently, by making it the subject of a precisely contradictory theory.

179. Really, the brainless thing has been happier than those who put it forward: for it certainly was innocent of any thought or ambition this way. Happier too is it than the luck- less children who have this unconscion-

able stuff put into their brains as "science," done up for them in their common school books. I know it is not politic for us to say so, at least within hearing of Dr. Buchner's school. To us who think that dosing a juvenile humanity with the bestial in thought, will probably elicit the bestial in action from the same humanity grown up—be- nighted as we are and stone-blind amid all this il- lumination of the mental doctrine as well as the moral process of the school—a man like Dr. Buch- ner flings back the word of utter contempt: " A howling pack of mental slaves, who deny that as the holothuria produced the snail, an ape or any other animal may have given birth to a man!" Exactly so! He says this in the preface to his *Kraft und Stoff*. Nothing truer! If you let your- self be spirited over one gap, there is no reason why you should not be spirited everywhere, up to man, or down to " the polar tensions of chemistry." In either case, having dispensed with reason in the process of logic, you are emancipated from moral- ity in the conduct of life; and that as well by the terms of your origin, as by the meaning of your

destiny. And the principles of your life, in your
short and checkered career from the cradle to the
grave, are best summed up in the words which the
Protestant Bishop of Carlisle quotes from a French
scientist, who exclaims with sentiment, if not with
taste: "Ah! is it not befitting to entertain a little
sane forbearance with respect to the seven capital
sins? Judge for yourselves. Just a little too much
blood, perhaps a hundredth part of a gramme ill-
directed upon contact with a little bit of nervous
fibre somewhere, and lo! on the spot a haughty
man, a vain woman, a proud creature!" Ah! happy
creature then, thou amphioxus, brainless, scaleless,
finless fish! Happier far without a brain, and with
a rudimentary heart, than hearts and brains of
men and women, with science and sentiments such
as these!

180. I have dwelt upon this one gap between the
mollusks and the vertebrates, as an instance of how
scientific men handle scientific facts.
Other Gaps.
Hurrying on to a conclusion, I will
barely mention a couple more of these chasms, un-
bridged in the geological records, but furnishing
me with a bridge good enough to approach our
concluding view.

181. In the cretaceous formation of North
America, which belongs to the secondary or meso-
zoic age, there occur leaves of the angiosperms,
which are plants of modern type, such as the wil-
low, elm, magnolia and the palms. They impart a

totally different character to forest vegetation from that of the preceding period. The same abrupt transition has been observed in Europe and other countries.

182. In the early tertiary, which was the beginning of the next age, called the cænozoic, the world was full of true mammals, many of great size, while no such mammal has **The Tertiary Age.** yet been detected in any earlier beds (No. 172 above). " The abruptness of the transition is astounding," says Professor Dana, "and needs facts for its full elucidation. The same abruptness in the introduction of the tertiary mammals occurs in the beds of other continents, as well tropical India, as colder Europe." This is the age that seems to correspond with the sixth day of the Mosaic cosmogony, where Moses places the production of the living creature in its kind, cattle and creeping things, and beasts of the earth according to their kinds.

183. Here, then, we are on the threshold of man's home, which is just being finished for his residence. It is not completed so far, that he can subsist in it as yet. Hence no one ever thinks of looking for man in tertiary times: they look for a kind of man, a progenitor to the one that is—according to evolution, a pithecoid man, an *anthropithèque*. He himself could not be. It was the time of great changes when whole genera of animate beings were undergoing modifications, by pro-

duction and extinction. All the orders were moving on, were carrying out the providential law of
development and progress through causes adequate
for their production, such as observation, logic and
philosophy can substantiate; not through an inade-
quate name or theory, to which no fact or law corresponds, and which every law and fact contradicts.

184. Here, upon the threshold of our own residence, you may turn round and look about you at
Nature as it is. nature as it is to-day. This will help
you to realize nature as it was then.
Among the orders and species developed as they
are, can you recognize any path by which evolution has travelled up? Where is the line, down
through the orders, down to the protoplasmic
moneron, or formless cell? Throw in all the hypothetical links and transitions which science can
hope to discover. Is there a line possible? Certainly, if evolution had no line of march, it never
marched. If it did march, where is the line, be it
straight, crooked or curved?

185. Mr. Herbert Spencer, following Lindley and
Professor Huxley, draws a diagram, in the second
part of his Principles of Biology, chapter 11; and
he endeavors to locate in a graphic way, all the
actual groups of animal nature, as they stand related to one another, and related to a common centre, protoplasm; for from this, on the evolutionary
theory, they should have evolved by some line or
other. The protoplasmic cell would thus be at

one end; man should come at the other; and all nature ought to lie between. How does the diagram look, as drawn by these scientific authorities?

186. The groups are dispersed towards all points of the compass, without any uniform angle of divergence from one another, and without any uniform distance from the common centre, protoplasm. Mammals are relegated into the distance like a far-off nebula in the sky. Worse than that: Mr. Spencer observes that no diagram on a plane surface can give any correct idea of the actual divergence, so irregularly scattered are the groups. "Such relations cannot be represented in space of two dimensions, but only in space of three dimensions." For the differences are so profound that "groups of the widest generality are based on characteristics of the greatest importance, physiologically considered." So that as to anything like a linear succession of groups below falling just a little short of groups respectively above them, by having their development arrested; and so on up to man; he says, referring to his diagram, "what remnant there may seem to be of linear succession in some of these sub-groups is simply an accident of typographical convenience. Each of them is to be regarded simply as a cluster."

187. The same holds with regard to plants. I shall simply quote some of his words. Speaking of the classification of the vegetable kingdom, he says, " here linear arrangement (that of a straight line by

which evolution might have marched) has disap-
peared; there is a breaking up into groups and sub-
groups and sub-sub-groups, which do not admit of
being placed in serial order, but only in divergent
and re-divergent order." So much for nature as
it is.

188. Possibly nature has changed her ways, and
become capricious in her old age. Has
she? What becomes of science if she
has? Science had better drop the subject and take to a more useful occupation. If the
combining properties of oxygen and hydrogen have
altered, and are no longer what they were; if the
melting point of platinum or the freezing point of
mercury was different 5,000 years ago from what it is
to-day; then possibly, and only then, has there been
a change in the laws of life. But if the chemical
element has a history from the time it first was, and
that history has been one of the precisest law, then
life, no less than chemistry, has its own history, one
of the precisest law, which we see in the present,
and which must have held in the past. Nay, what
else is the meaning of the talk about "the great
secular processes of the Darwinian laws," if laws
are not laws, and if caprice is to dominate over nature? "If the simplest forms of the present and
the past were not governed," asks Dr. Dallinger,
"by accurate and unchanging laws of life, how did
the rigid certainties that manifestly and admittedly
govern the more complex and the most complex

Nature as it was.

come into play ? " Once establish," he says, " by clear and unmistakable demonstration, the life-history of an organism, and truly some change must have come over nature as a whole, if that life-history be not the same to-morrow as to-day; and the same to one observer, in the same conditions, as to another." " Every piece of living protoplasm we see has a history: it is the inheritor of countless millions of years. It is the protoplasm of some definite form of life, which has inherited its specific history. It can no more be false to that inheritance, than an atom of oxygen can be false to its properties." Both have been as they are from the beginning. Like the lines in the solar spectrum, they are parallel to all other lines; and forms may come in between, only to be parallel still in their history. They may group together in certain general colors as it were, in the yellow, and the red and the blue; but they are parallel everywhere; and never meet. Where, then, did these lines of species come from ? An adequate cause. What is that ? Not transformation of one into another.

189. We will admit therefore what is correct in the statement of evolution. We will tolerate the expression of it even in such turgid and loose declamation, as that " the irrefrag- **The Facts of Progress.** able philosophy of modern biology is that the most complex forms of living creatures have derived their splendid complexity and adaptations from the slow and majestically progressive

variation and survival from the simpler and the simplest forms." We will only challenge this phraseology so far as to discount for its looseness and irregularity. There is not much harm in letting these phrases go. There is policy in it. And I notice that the paroxysm of admiration for "the secular processes of the Darwinian laws" lasts only awhile. *Natura usque redibit,* nature, and just common sense return to assert themselves. For "the Darwinian laws, by the way," adds the same writer, Dr. Dallinger, in the same place, " could not operate at all, if caprice formed any part of the activities of nature."

190. The amount of truth to be admitted is this. There has been progress in nature. In the first place, the whole development of this

1. From the
Simple to the
Complex.
Haeckel's
Phylogenesis.

earth, and of all the life upon it, has proceeded on the plan of beginning in the simple sea-plant or lower forms of animals, and ending in man; beginning with even an embryonic simplicity, and, like an embryo that proceeds from the simple to the complex, advancing to a general prevalence of complex organizations over the world.

191. This fact so stated is what suggested to Professor Haeckel his argument of *phylogenesis,* according to which, every embryo represents, in the stages of its development, the different species through which its ancestors passed, from the original formless cell up to its present specific form. The

parallelism not being perfect, he supplied what was wanting. At his own risk and expense, he lent a few " links" to nature, and completed an imaginary line of her species in history. Finding that, even so, the line in nature's species was not what it should be to match the embryo's line of development, and was not even a line at all, he introduced into nature a " law of falsifications," whereby she plays him false. That he calls *cenogenesis*. Then he complements this, of course, with a " law of verifications," whereby she is honest with him. This he calls *palingenesis*. But even so the argument cannot be made to stand, nor his evolution to keep steady on the top of it; so he slips in a practical law of his own personal falsifications, whereby he fabricates in his engravings certain elements to build up his theory anyhow. Now, what is called Science stood everything up to this. But, as great men of his celebrity have a little rivalry and criticism playing upon them, his critics and his rivals in the field could not resist the temptation here of exposing him baldly. It was the only grievous sin he committed,—to fabricate a few plates. To rail at God was nothing, that was only an offence against piety. But to tamper with an engraving was an insult to their understanding. So the argument of his genius has become, for the nonce, the subject of " an arrested development."

192. To continue the statement of the facts:—it was, in the second place, a progress in climate and

other conditions that involved a concurrent progress

2. Conditions of the Progress. from the inferior living species to the superior. Coming down from the Silurian age to the present, through those epochs and periods which I had occasion to sketch before (No. 31), there has been a more comprehensive succession of phases in the life of the world, than there is to-day between the equator and the poles, though that range varies in its climate from the tropical and subtropical zones, through the temperate, subarctic, and arctic regions. There was much more in the progress of the world.

193. Hence, thirdly, there was an arithmetic of progress about it; though one far different from that

3. Its Arithmetic. which evolution would require. If animal species evolved from one another, we should begin with one, or with a small number, and reach by successive stages of geometrical progression the present ample quantity of nearly 150,000. Suppose we began with 10, in the Silurian; and that amounted to 34 in the Devonian, and to 111 in the Carboniferous, and 387 in the Permian: then in the secondary formations, proceeding at the same rate of geometrical progression, we should have in the Triassic, Jurassic and Cretaceous, 1163, 3830, 12,614, respectively: in all the tertiary formations, let us take only one step further, and put it down at 45,500 species; then, in the present, at the same rate of progress we should arrive at the actual facts of the case, and find the 150,000 species of animals. This process we suggest in the interests of evolution, to show its way in the world. But unfortunately the best will cannot save it, as the world goes; the world itself turns round and will not have it. For what are the facts of the case, as they stood just recently? Though many new species are found, yet the proportions of the following table need not be considered changed. Instead of beginning with 10, and going on as we have reckoned, we start and proceed in the following ratios: 10,209, 5160, 4901, 303, 1310, 4730, 5500, 16,970, and 150,000 in the present age. The arithmetic is entirely out of order and false, if evolution is true.

194. Where did the 10,209 of the first age come from, or the 150,000 of the present? Where have their ancestors gone to? What fatality descended on nature as it was, if nature as it is comes down from it by the descent of species?

195. As to the question whence the 10,209 of the first or Silurian age came, Mr. Darwin answers as we quoted him before (No. 66). He says, in his Origin of Species: "Where are the remains of those infinitely numerous organisms which must have existed long before the first bed of the Silurian system was deposited?" Mark the words, "which must have existed"! He answers the question thus: "They may all be in a metamorphosed condition, or may be buried in the ocean."

196. And fourthly, as to anything like a chain of

evolution, which should have given us by regular
progression the present world of species,
he observes a little earlier: "Geology
surely does not reveal any such finely
graduated chain." But I will express this more
satisfactorily in the terms of the French scientist,
M. Gaudry, director of the Museum, and a decided
evolutionist, who says in his Primary Fossils: "The
most able observers refuse to admit a single linear
series, beginning at the monad, continuing in due
course under the form of polyp, echinoderm, mollusk, annelid, articulate, fish, reptile, bird, mammal,
and finishing in man. Although the mammals are
the most perfected of the vertebrates, the study of
their embryonic development does not show us that
they ever passed through the fish state, and through
the bird state. Palæontology here confirms em-
bryology, when it considers itself to have discovered in geological times, that there was not any
single chain of beings, but many such chains, the
development of which has gone on independently."

4. No Single Chain of Beings.

197. Now, we are on the threshold of man's
household. But we must not cross it. Biology, all
about life in general, does not cover the
psychology, which is specially about the
life of man. We may look wistfully
towards him; we may mount, if you like,
on the shoulders of those orders, which seem to the
casual observer so provokingly similar to man—the
monkeys, the apes. But we have reason to fear

Sixth Day: Second Part. Psychology.

that, whether biologically considered, or psychologically, or practically, the shoulders of the same apes will be found, by even the best logicians, to be an unsafe post for taking observations. We may show them all deference, to soften the asperity of their temper or of their finger nails in such a delicate contingency; take account of their foibles, as Mr. Darwin has done; soothe them with Darwinian compliments, that they are so like man; forasmuch as, according to Mr. Darwin, they are fond of tea and coffee and sugar; and they do not disdain tobacco, beer, and spirituous liquors! We may compassionate them, that alas! they get headaches in consequence of their indulgence therein, just as their betters do, while—judicious creatures!—they forego all such indulgence for the future, as some of their betters do not. We may admit that the good, sweet brutes get sick with pulmonary catarrh, consumption, apoplexy, intestinal inflammation, cataract of the eyes: yes, and that medicine can cure them. What then? Mr. Darwin has tried all this, and what has been the consequence? Simply that it has been judged safer to get off the brutes' backs and leave the apes alone. Vogt has told us already (No. 138) that the lowest apes are too far gone in evolution to submit to any such operation now. Indeed, long ago mankind knew these things, and took account of these analogies between the brutes and man. Any casual observer can see them for himself, or might readily suspect them.

And what did mankind conclude?   Not that man
was descended from the ape.   But that the ape was
an animal; and so was man.   Only he was a rational
animal.   It is in the rationality of the thing, as in
the rationality of the argument, that the difference
comes in between evolution and the common sense
of mankind.

198. Here then we may pause, considering ourselves happy in being dispensed by the obvious facts
of human life from sinking man into the mists of a
materialistic biology.   He has a sphere of his own,
an intellectual atmosphere, to move about and
breathe in; and we have a right to respect it—his
rational psychology.   In the steaming valley of
sense he is awhile; but his home is not here.   He
is in it awhile on sufferance, by the law of his
organic nature; but he is not of it,—by the higher
law of his spiritual being and its destiny.   Those
who like the steaming exhalation which greets the
sense in the valley of matter, and deadens the intellectual life, are free to enjoy it.   We may prefer to
have none of it.

199. Yet I entertain a hope even for them; and
certainly for the outcome of all science of theirs,
however much it is misinterpreted in its earlier and
cruder efforts.   If it is only fact which is reported,
and law which is rendered to the inquiring mind,
the very mists of a sensualistic science, which
obscure the vision and oppress the soul, can yet be
lifted into the broad light of God's open sky;

where, tinged and painted as they rise, they weave themselves into a texture of gold, cinctured with bands of watered satin, reflecting the rays of His Providence. The cloud which threatened to hang over us, like a pall of intellectual death, becomes but the light summer vapor, which hangs pendent over the firmament on a smiling day. And draping, as in a beautiful tapestry, the broad azure of one divine conception over the world, such a floating tribute of human science to the truth of things above is but an adornment to the golden sun of God's Providence, which like the eye of heaven beams benignly down, with the surpassing glories of His love. These things we may hope for, as science and sense prevail. And, basking in the genial hope as in a sunny dream of the future, we may salute it coming—all efforts of human genius we salute at every stage—

> Ye mists and exhalations, that now rise
> From hill or steaming lake, dusky or gray,
> Till the sun paint your fleecy skirts with gold,
> In honor to the world's great Author rise!
> Whether to deck with clouds the uncolored sky,
> Or wet the thirsty earth with falling showers,
> Rising or falling still advance His praise!

---

200. While dedicating these pages to the cultured classes and students of the community, I would beg to observe at the same time what several points there are, for which these pages are not responsi-

ble.   To be supposed to have taken up what has
not had justice done to it, or to have failed in doing justice to what was taken up, would alike be
out of keeping with the respect due to them, and
with the real gravity of the subject.   The constructive view of Life, or what is the Vital Condition,
whether that belongs to biology or to psychology, has
not been treated here.   Again, the several systems
or explanations of Progress or Evolution which are
in accord with the facts of science, and are no less
in accord with sound sense and logic, belong to a
constructive and synthetic essay on the subject;
not to this critical analysis of prevalent theories,
which has here been broached.   The spirit that
has animated the criticism might appear somewhat
destructive; still I trust it has been one of candor
and strict truth.   For certainly it is the exigency
of the situation, not a predilection for the process,
which has engaged us all in this precise phase of
philosophical criticism.

L. D. S.

PRINTED BY BENZIGER BROTHERS, NEW YORK.

www.ingramcontent.com/pod-product-compliance
Lightning Source LLC
Chambersburg PA
CBHW021804190326
41518CB00007B/444